A Nordic Smart Sustainable City

This book critically explores research and development on the smart sustainable city, emphasizing the tension and association between smartness and sustainability, both as a concept and as a phenomenon in a Nordic context.

Worldwide, increasing urbanization and its related challenges, along with urgent environmental issues, have sped up the international interest for smart, sustainable cities as a concept that could increase the efficiency of services, minimize environmental impacts, and improve the quality of living in cities and urban areas. This book scientifically discusses the provenance, substance, and processes of the smart sustainable city, with illustrative examples of how it is translated into urban realities in a medium-sized city, drawing upon Stavanger, one of the first, and one of the leading smart sustainable cities in Europe. The book's multi-disciplinary perspectives and thematic lenses include education and knowledge, arts and culture, safety, climate and sustainability, mobility and transport, economics, democracy, participation, innovation and entrepreneurship, data, and communication. While demonstrating the academic breadth and wide-ranging impact of the smart sustainable city concept, the book promotes and updates the ground for mutual understanding, communication, and collaboration between multiple disciplines and stakeholders involved in developing functional, democratic, and sustainable solutions for the urban present and future.

A Nordic Smart Sustainable City: Lessons from Theory and Practice presents an overview of scientific and practical current approaches in a readable format for practitioners and administrators in municipalities and related businesses, for researchers, academics, educators, students, and stakeholders.

Barbara Maria Sageidet is a professor of natural science in the Department of Early Childhood Teacher Education at the University of Stavanger (UiS). She has a PhD in soil and environmental sciences from the Norwegian University of Life Sciences (NMBU), Ås, related to paleoecology and soil micromorphology. Her research relates to natural science, natural science didactics and sustainability in kindergarten, and environmental citizenship, with interests in environmental and soil literacy, urban gardens, and urban childhood.

Daniela Müller-Eie is a professor of city and regional planning in the Department of Safety, Economics, and Planning at the University of Stavanger and holds a PhD in architecture/urban sustainability from Glasgow University. Her research generally focuses on the interaction between the physical environment, planning measures, socio-cultural conditions, and psychological factors. More specifically, she studies sustainable urban mobility and travel behaviour and related incentives.

Kristiane M.F. Lindland is a research manager for climate, environment and sustainability in the division for Health and Society at NORCE Research and holds a minor position as an associate professor in change management in the Department of Media and Social Sciences at the University of Stavanger. She holds a PhD in management from the University of Stavanger. Her research areas stretch from innovation, design, leadership, and organization to energy justice, citizen involvement, and sustainability. What characterizes her approach to these themes is a relational and processual understanding of reality.

Routledge Studies in Sustainable Development

This series uniquely brings together original and cutting-edge research on sustainable development. The books in this series tackle difficult and important issues in sustainable development including: values and ethics; sustainability in higher education; climate compatible development; resilience; capitalism and de-growth; sustainable urban development; gender and participation; and well-being.

Drawing on a wide range of disciplines, the series promotes interdisciplinary research for an international readership. The series was recommended in the Guardian's suggested reads on development and the environment.

Good Education in a Fragile World
The Value of a Collaborative and Contextualised Approach to Sustainability in Higher Education
Edited by Alan Bainbridge and Nicola Kemp

Greening Higher Education in Europe
Institutional Transitions to Sustainable Development
Magdalena Popowska

Implementing Sustainable Cities
Edited by Sylvie Albert, Jeremy Millard, and Manish Pandey

Sustainable Urban Development in the European Arctic
Dorothea Wehrmann, Michał Łuszczuk, Katarzyna Radzik-Maruszak, Jacqueline Götze, and Arne Riedel

A Nordic Smart Sustainable City
Lessons from Theory and Practice
Barbara Maria Sageidet, Daniela Müller-Eie, and Kristiane M.F. Lindland

For more information about this series, please visit: www.routledge.com/Routledge-Studies-in-Sustainable-Development/book-series/RSSD

A Nordic Smart Sustainable City

Lessons from Theory and Practice

Edited by
Barbara Maria Sageidet,
Daniela Müller-Eie, and
Kristiane M.F. Lindland

First published 2025
by Routledge
4 Park Square, Milton Park, Abingdon, Oxon OX14 4RN

and by Routledge
605 Third Avenue, New York, NY 10158

Routledge is an imprint of the Taylor & Francis Group, an informa business

British Library Cataloguing-in-Publication Data
A catalogue record for this book is available from the British Library

ISBN: 978-1-032-81211-3 (hbk)
ISBN: 978-1-032-81212-0 (pbk)
ISBN: 978-1-003-49865-0 (ebk)

DOI: 10.4324/9781003498650

Typeset in Galliard
by Taylor & Francis Books

Contents

PART IV
Lessons learned   217

About the Editors

Barbara Maria Sageidet is a professor of natural science in the Department of Early Childhood Teacher Education at the University of Stavanger (UiS). She has a PhD in soil and environmental sciences from the Norwegian University of Life Sciences (NMBU), Ås, related to paleoecology and soil micromorphology. Her research relates to natural science, natural science didactics and sustainability in kindergarten, and environmental citizenship, with interests in environmental and soil literacy, urban gardens, and urban childhood. Since 2016, Sageidet has been part of the emerging network at the UiS related to smart sustainable cities, and since 2019 she has been an active member of the Research Network for Smart Sustainable Cities at the UiS. Sageidet is a member of the Filiorum (UiS) and has also worked within the KINDknow (HVL), both Norwegian research centers on early childhood education. She has reviewed for international research journals within science education, early childhood (teacher) education, and environmental sciences.

Daniela Müller-Eie is a professor of city and regional planning in the Department of Safety, Economics, and Planning at the University of Stavanger and holds a PhD in architecture/urban sustainability from Glasgow University. Her research generally focuses on the interaction between the physical environment, planning measures, socio-cultural conditions, and psychological factors. More specifically, she studies sustainable urban mobility and travel behaviour and related incentives. Müller-Eie is an active member of the Research Network for Smart Sustainable Cities at the UiS. She has experience in leading and participating in national and international research collaborations and has reviewed for several international research journals.

Kristiane M.F. Lindland is a research manager for climate, environment and sustainability in the division for Health and Society at NORCE Research, and holds a minor position as an associate professor in change management in the Department of Media and Social Sciences at the University of Stavanger. She holds a PhD in management from the University of Stavanger. Her research areas stretch from innovation, design, leadership, and organization to energy justice, citizen involvement, and sustainability. What characterizes her approach to these themes is a relational and processual understanding of reality. She has led and participated in several research projects, both at national and EU-level, with collaborators across various disciplines and in close collaboration with participating citizens and public collaborators. Lindland is an active member of the Research Network for Smart Sustainable Cities at the UiS. She has reviewed several articles and book chapters.

Contributors

Kristoffer Berre Alberts in an associate professor at the Department of Classical Music, University of Stavanger, Stavanger, Norway. He finished his artistic research PhD in 2023, a project that moved away from direct musical interaction, and into a landscape and tradition taken from noise music, where he experimented with compositions of isolated and soloistic sonic elements. Alberts has made an international impact as a musician in the landscape of musical improvisation. He has collaborated with musical personalities such as Nels Cline, Okkyung Lee, Maja Ratkje, Steve Noble, Jamie Saft, Terrie Ex, Paal Nilssen-Love, Han Bennink, and Lasse Marhaug. He runs the record label KBA Records and appears as a saxophonist on eleven record releases on the Portuguese record label Clean Feed.

Helga Aunemo holds a M.Sc. in globalization, with an emphasis on global politics and culture. Based in Brussels since 2016, she is following EU initiatives where they develop. Since 2021, she has worked as an EU advisor for the Stavanger Region European Office, Brussels, Belgia, where she has been leading the region's strategic efforts within active and healthy ageing at European level. She is passionate about sustainable urban development, and the opportunities the twin transitions hold for local communities.

Helene Eiliott, European advisor at Stavanger Region European Office in Brussels, holds a M.Sc. in science and technology studies, with an emphasis on innovation and strategy (2008). As self-employed and working in the creative sector in Stavanger, she co-founded and managed an artist-run gallery, Prosjektrom Normanns, and a co-working space, Elefant. Since 2018, she has worked for the municipality of Stavanger, being responsible for the smart city program for innovation and use of artistic research in urban development and societal transitions. Her latest project on behalf of the municipality was the EU-funded lighthouse project for the New European Bauhaus movement: NEB-STAR, where 16 partners explore how municipal plans can be a societal tool to accelerate the green transition.

Petter Frost Fadnes is a Norwegian saxophone player, lecturer, and researcher based at the University of Stavanger, Stavanger, Norway. With a PhD in performance from the University of Leeds (2004), Frost Fadnes was for many years part of the highly creative Leeds music scene, and now performs regularly with The Geordie Approach, Mole, and Kitchen Orchestra. Frost Fadnes has published on a wide range of performance-related topics, such as jazz collectives, cultural factories and jazz places, film scoring, jazz for young people, and improvisational practices. He is professor of improvised music at the Faculty of Performing Arts, and former principal investigator for the HERA-funded research project Rhythm Changes: Jazz Cultures and European Identities. His book, *Jazz on the Line – Improvisation in Practice*, was published at Routledge in 2020.

Jens Kaae Fisker is an associate professor of social science in the Department of Media and Social Sciences at the University of Stavanger, Stavanger, Norway. With a background in geography, he holds a PhD from the Department of Planning and Development at Aalborg University. His work sits at the intersection of urban, rural, and regional studies as seen from the perspective of political geography. Jens is currently involved in the New European Bauhaus project NEB-STAR and is a founding member of the Social and Spatial Justice Research Group at the University of Stavanger.

Kenneth Pettersen Gould is associate professor in risk management and societal safety at the Department of Safety, Economics, and Planning, University of Stavanger, Stavanger, Norway. With a specializing in organizational risk and safety management, his research concerns hazardous technologies, including how risk, reliability, safety, and security is analyzed and managed in organizational contexts, and how regulatory and management strategies in these areas can be further developed.

Stella Huang is a PhD candidate in technology and social change at the Department of Thematic Study (TEMA), Linköping University, Linköping, Sweden. At the time of writing this chapter, she is serving as a lecturer in the Department of Media, Culture, and Social Science at the University of Stavanger. Her current research project centers on advancing knowledge of data-driven tools to support urban governance efforts towards sustainability and climate neutrality. She has authored peer-review articles in international journals, contributed several chapters to international books in geopolitics and sustainability, and co-edited the book *Dilemmas, Contradictions and Paradoxes in Sustainability Thinking*.

Ayda Joudavi is a PhD candidate in city and regional planning at the Department of Safety, Economics, and Planning, University of Stavanger, Stavanger, Norway. She has been a visiting PhD scholar at McCormick School of Engineering and Northwestern University in Chicago, USA. With an academic background in architecture and urban planning, she collaborates with an EU-funded research project called "Greencoin," which aims to promote pro-environmental behaviours through a reward-based system and gamification. In her research, she studies the role of incentives in promoting commute cycling and sustaining travel behaviour change towards sustainable modes of transport.

Todor Milkov Kesarovski is a PhD candidate at the Department of Safety, Economics, and Planning at the University of Stavanger, Stavanger, Norway. He currently develops the research project "Optimising Urban Densities for Sustainable Cities". In his work, he explores the potential of the concept of density to support advances in the modelling of urban environments and evaluating territorial balances within metropolitan areas through big data and GIS analytical techniques.

Oluf Langhelle is professor in political science at the Department of Media and Social Sciences, University of Stavanger, Stavanger, Norway. He earned his Dr. Polit. degree at the University of Oslo, Norway. His research has focused on the concept of sustainable development and follow-up, strategies for sustainable development, environmental politics, and policy, including oil and gas policies in the Arctic, transitions towards low carbon societies, focusing on carbon capture and storage (CCS), and the Doha Round Negotiations in the World Trade Organization (WTO).

Veronika Lorentzen is a PhD candidate in sustainability transitions leadership at the Department of Media and Social Sciences, University of Stavanger, Stavanger, Norway. She is researching affordable and sustainable housing options and housing politics in Norway. She has been a participant in the regional final of the Norwegian Researcher Grand Prix 2024 with her research project on Svartlamon experimental area in Trondheim. Veronika holds three master's degrees: one in event management, one in leadership, and one in fashion design, all unified by a strong interest in sustainability.

Fredrik A.F. Matre, Public Policy Projects, London, United Kingdom, is an early-career researcher with a bachelor's degree in international politics from Aberystwyth University, UK, and a master's degree in politics and international studies from the University of Cambridge, UK. His research interests are focused on climate change, epistemic communities, and relational approaches to international relations theorizing, including social quantum theory and pragmatism. He has previously worked in electoral services for local government and for an environmental INGO. He currently works at the health and life sciences think tank Public Policy Projects.

Anders Riel Müller is an associate professor of city and regional planning at the Department of Safety, Economics, and Planning at the University of Stavanger, Stavanger, Norway. His main research interest is to understand how imaginaries of progress shape political, economic, and cultural relations of production and consumption with a particular focus on aspects of justice in relation to urban planning, land use, energy transitions, and food systems. He is currently leading the Research Network for Smart Sustainable Cities at the UiS, and the research and innovation project Future Energy Hub funded by the Research Council of Norway.

Gianfranco Nencioni is an associate professor at the Department of Electrical Engineering and Computer Science, University of Stavanger, Stavanger, Norway, since 2018. He has received the M.Sc. degree in telecommunication engineering and the PhD degree in information engineering from the University of Pisa, Italy, in 2008 and 2012, respectively. In 2011, he was a visiting PhD student at the Computer Laboratory, University of Cambridge, UK. He was a post-doctoral fellow at the University of Pisa from 2012 to 2015 and at the Norwegian University of Science and Technology, Norway, from 2015 to 2018. He is currently the head of the Computer Networks (ComNet) research group and leader of the project 5G-MODaNeI funded by the Norwegian Research Council. His research activity focuses on modelling and optimization in emerging networking technologies (e.g., SDN, NFV, 5G/6G, network slicing, multi-access edge computing). His past research activity had a focus on energy-aware routing and design in both wired and wireless networks and on dependability of SDN and NFV.

Chandra Prakash Paneru is a PhD student at the Department of Safety, Economics, and Planning, University of Stavanger, Stavanger, Norway. He is a member of the Greencoin project. His research focuses on promoting smart and sustainable energy practices. He earned a bachelor's degree in civil engineering from Tribhuvan University (Pokhara, Nepal) in 2011 and pursued a master's degree in energy and environmental technology from Telemark University College in 2014, followed by three years' of work experience. Returning to academia, he earned a second master's degree in city and regional planning at the University of Stavanger in 2020. Chandra has a strong interest in the interdisciplinary fields of urban planning, sustainable energy, smart innovation, and pro-environmental behavior.

Harald N. Røstvik is a professor emeritus in city and regional planning at the Department of Safety, Economics, and Planning, University of Stavanger, Stavanger, Norway. As an Architect MNAL from the School of Architecture, University of Manchester, UK, he has been a pioneer and a visionary of sustainable design. From 1977 to 2015 he designed and researched sustainable buildings, communities, and transportation systems in European, African, and Asian climates through his practice SunLab. He has been teaching city- and regional planning-related sustainability at the University of Stavanger since 2015. From 2010 to 2015, he held the professorship of durable design at Bergen School of Architecture, while still in private practice. He has written eight books.

Annisa Sarah is a PhD candidate at the Department of Electrical Engineering and Computer Science, University of Stavanger, Stavanger, Norway. She is specializing in resource allocation within the 5G multi-access edge computing ecosystem. She holds an M.Sc. in wireless systems from KTH Royal Institute of Technology, Sweden (2017) and a BSc in telecommunication engineering from Telkom University, Indonesia (2014). Annisa previously served as an assistant professor in the electrical engineering program at Atma Jaya Catholic University of Indonesia (2018–2021), where she focused on teaching and research in telecommunications. Her research interests include wireless networks, network optimization, and rural connectivity solutions.

Ruth Østgaard Skotnes is an associate professor in risk management and societal safety at the Department of Safety, Economics, and Planning, University of Stavanger, Norway. She has a PhD in social science with specialization in risk management and societal safety from the University of Stavanger, Norway, and a M.Sc. in administration and organization science from the University of Bergen, Norway. Her research interests include critical infrastructure protection, vulnerability and technology, cyber safety and security, and cyber-physical systems. She has led and participated in several research projects on safety and security in critical infrastructures, risk communication, risk regulation, digitalization and cybersecurity, and safety and security culture.

Reidar Staupe is an associate professor at the Department of Safety, Economics, and Planning, University of Stavanger, Stavanger, Norway, and at the Department of Technology and Safety, at the Arctic University of Norway, Tromsø, Norway. Notable publications include the 2021 monography *Disasters and Life in Anticipation of Slow Calamity Perspectives from the Colombian Andes and the upcoming volume A time of disastrous anticipations: Essays on life in the shadow of catastrophe*, both published by Routledge. His research interests revolve around disaster temporalities and lived experience.

Helleik Rosenvinge Syse is a PhD candidate at the Department of Energy and Petroleum Engineering, at the University of Stavanger, Stavanger, Norway, researching at the University of Stavanger's Future Energy Hub. His work explores the drivers of building energy consumption, energy efficiency improvements, and local energy production solutions. In 2023/24, he was a visiting scholar at the University of New South Wales's AI Institute in Sydney, Australia. Helleik specializes in the techno-economic analysis of renewable energy and building energy systems, focusing on the intersection of engineering and human behavior. He earned his M.Sc. in renewable energy systems and the environment from the University of Strathclyde, UK, in 2016 and has worked in multiple startups and scale-ups. As an educator, he has taught sustainability-focused courses at BI Business School. Recognized as a "30 under 30" by Dagens Næringsliv (2020) and a "young energy talent" by the Stavanger Chamber of Commerce (2022), Helleik also co-hosted the podcast *Energisnakk*, discussing energy, environment, and climate challenges.

Ari K.M. Tarigan is an associate professor at the Department of Safety, Economics, and Planning, at the University of Stavanger, Stavanger, Norway. He earned his PhD in transportation from the Graduate School of Engineering, Kyoto University, Japan, and his M.Sc. in municipal water and infrastructure from the IHE Delft, Netherlands. His research interests include road safety, traffic engineering, travel behavior, transportation planning, and urban infrastructure. He has recently been involved in projects such as promoting pro-environmental behaviors among urban citizens, nature-based solutions, stormwater management, and green infrastructure.

Ståle Undheim is the research station manager at the Norwegian Institute of Bioeconomy Research (NIBIO), in Klepp, Norway. NIBIO is one of the largest institutes in Norway contributing to food safety, sustainable resource management, innovation, and value creation. He came to NIBIO from a position as manager of The Smart City department in Sola municipality. He has 30 years' of work and organizational experience from various positions in local and regional administrations and 13 years as a chair of the board of the Norwegian Association of Municipal Engineers. Undheim has a master's degree in change management from the University of Stavanger, Stavanger, Norway, and has studied related subjects at various other Norwegian universities. His competency profile is sustainability, innovation, planning, and management.

Petar Zhivkov is a PhD candidate at the Department of Information Technology and Communications at the Bulgarian Academy of Sciences, Sofia, Bulgaria, where he develops the research project "Analyzing the Impact of Atmospheric Air Quality on Quality of Life". In his work, he explores the potential to support advances in the modelling of urban environments, calibrating air quality data, and enhance strategies for protecting humans' health through big data, machine learning, and GIS analytical techniques.

Preface

This book has evolved out of many discussions about the importance of a common cross-disciplinary publication on smart city research in Stavanger in the Research Network for Smart and Sustainable Cities at the University of Stavanger. This outcome is the result of a long process and committed writing and has been a journey of learning and development, collegiality and friendship, collaborative explorations and experiences. We are all immensely grateful to Anders Riel Müller, the research network leader, who made this possible.

Very much thanks to all who have inspired, facilitated, encouraged, and promoted this opportunity in many ways: The Stavanger region-wide smart city network members from the Stavanger Municipality; the neighbouring municipalities Sandnes, Sola, and Gjesdal; and Nordic Edge, the Norwegian Research Centre NORCE, Lyse, Forus Næringspark, Kolumbus, Universitetsfondet, and the University of Stavanger for valuable scientific, practice-oriented, financial, and personal collaborations and support.

Personal thanks are due to many! Multiple essential persons know the roles they have had, and we do! Among them are Troels Gyde Jacobsen, Gunnar Edvin Crawford, Trygve Meyer, Terje Eide, Karl Fjelde Nevland, Kristin Kverneland, Emilie Christensen, Heidi Kristina Jakobsen, Chunming Rong, Harald Sævareid, Tegg Westbrook, Fabio Palacio Hernandez, Tord Tordbjørensen, Dagfinn Vaage, Tone Grindland, Anne Lise Falch Anfinsen, Mette Fossan, Gerd Seehus, Jennifer Clark, Jung Won Sonn, Gabriele Brennhaugen, Finn Arne Jørgensen, Runa Monstad, and Øystein Lund Bø.

We also want to thank everybody who has contributed to the data production and gathering for the various chapters and all practitioners, informants, and stakeholders who have contributed to realizing this book through practical help, conversations, interviews, narratives, and observations. Especially, we want to thank all the dedicated chapter reviewers for their valuable scientific input through several rounds! We greatly appreciate their help, which has been essential to achieving the scientific standard we aimed for. We also want to thank the reviewers of the book proposal and the first draft of the publication. Thanks also to Rosie Anderson and Matthew Shobbrook and others at Routledge for their practical help and essential support throughout the process.

The publication of this book is financially supported by the University of Stavanger, Nordic Edge, and Universitetsfondet.

We further want to thank all contributors' affiliations, among them departments and centres at the University of Stavanger, across faculties, including Filiorum (Centre for Research in Early Childhood Education and Care), for collegial, financial, and other kinds of valuable support. Finally, we want to thank all the people around us who have supported us in so many ways.

Warm and cheered thanks to all!

Stavanger, October 2024
Barbara Maria Sageidet
Daniela Müller-Eie
Kristiane M.F. Lindland
All chapter authors

Introduction

Barbara Maria Sageidet, Daniela Müller-Eie, and Kristiane M.F. Lindland

Introduction and background

In response to global urbanization and related environmental, social, and economic challenges, new urban concepts have emerged to address increasingly complex systems such as cities and their infrastructures, among them the smart city (Dameri, 2013; de Jong et al., 2015; UNHABITAT, 2022). Cities and municipalities around the globe acknowledge the importance and necessity of technology and knowledge, including the related sociocultural and intellectual capital and citizen participation (Albino et al., 2015; Cizelj & Sinkovec, 2012; Keshavarzi et al., 2021; Sageidet & Heggen, 2021; Zhao et al., 2021). Also, cities realize the potential of these resources on the way towards a more sustainable urban future (Bibri & Krogstie, 2017).

The smart city agenda gradually emerged a few decades ago, both in America and in Europe (Fairfield, 1994; Shelton et al., 2015). In Europe, the smart city agenda has been proliferated by European Commission Framework programs like Horizon 2020 (Späth & Knieling, 2020) and Horizon Europe towards 2030, including the EU mission: climate-neutral and smart cities (Gaglione, 2023). *Smart city* concepts have become widely applied in local governments, business development, and urban strategies, as well as in the context of urban research and development (Batty, 2017). The emergency of global environmental challenges has increasingly intertwined the smart city concept with the global Sustainable Development Goals (United Nations, 2015), especially Goal 11. Despite multiple ongoing tensions, related processes have in many contexts driven forward a gradual transformation of the smart city concept into the *smart sustainable city* concept (Bibri, 2019; Duygan et al., 2022). Forced by these proliferous applications, there are still multiple open questions about the definition, conceptualization, content, and relevance of the smart city concept, particularly from an interdisciplinary point of view; that is, when considering the many ways in which the concept can be understood, applied, and integrated in various fields (Dameri, 2013). There are also clear discrepancies between the smart city as a vision and its empirical realities.

The Research Network for Smart Sustainable Cities at the University of Stavanger (UiS) is an inter-faculty initiative, involving researchers from the humanities, educational sciences, social sciences, and technical and natural sciences. The

DOI: 10.4324/9781003498650-1

research network realized that multi- and interdisciplinarity bring together distinctly different research areas, using different terminologies, conceptions, methods, and theories – which can sometimes lead to misunderstandings and tensions (cf. Derry et al., 2014; Menken & Keestra, 2020). Thus, the collaboration of researchers from disciplines across the network, including collaboration with external partners, has been an exciting and constructive process for the researchers at the UiS, encompassing experiences of development, both collectively and individually. The research network underlines the need for a multiple-perspectives approach to urban sustainable futures, and it intends to strengthen interdisciplinary research and cross-sectional collaboration to meet the complex challenges in a smart city.

This book is a collection of chapters on theoretical and conceptual views of the smart city across disciplines and fields, as well as practical cases describing concrete applications and realities of the smart sustainable city 'case' Stavanger. Stavanger is a medium-sized city and municipality in Norway, with approximately 148,000 inhabitants. With a history of fishery and shipping industry until the early 20th century, engineering industries, mostly related to the offshore petroleum industry, have dominated since the 1970s. Within the EU Smart City lighthouse program, Stavanger became a smart city in 2014 (Fernandez et al., 2023), and it is currently one of 112 European cities that aims to be climate-neutral by 2030 (Gaglione, 2023).

Based on the Nordic medium-sized city of Stavanger, the book will illustrate, share, and reflect on multiple perspectives of research and practice in a lived smart city, ranging from theory and conceptual discussions to operationalizations, effective assessment, and case studies, while also addressing sustainability. The contributions are from academics and practitioners and thus provide a unique insight into a Nordic Sustainable Smart City.

Aim of this publication

This book is the result of collaborations between members of the Research Network for Smart Sustainable Cities at the UiS and academic, public, and private contributors. It aims to intervene with critical input and alternative perspectives in the following ways:

- Providing holistic views on the smart city concept through inter-, multi-, and trans-disciplinary perspectives across science and technology, social sciences, and the humanities.
- Exploring, critically reflecting, and clarifying the merging, possible synergies, dichotomies, and paradoxes of smartness and sustainability.
- Investigating particularities of a Nordic medium-sized smart city aiming to be climate-neutral by 2030.
- Critically exploring various theoretical perspectives and positions on the smart sustainable city concept within several disciplines.
- Actively describing, reflecting on, and using practical and concrete implementations of the smart sustainable city concept.

Each chapter in this book addresses at least one of these aims and may touch upon several existing strands of discourse concerning smart sustainable cities.

Holistic and inter-, multi-, and trans-disciplinary

Internationally, there has been an evolution from technology-focused approaches of the smart city towards a more holistic perspective, integrating governance, social capital, urban development, people, and community needs (Albino et al., 2015; Keshavarzi et al., 2021). How different disciplines define and conceptualize smart city concepts is interesting because it strongly influences how we discuss and approach the concept and its relevance. Viewing the smart city concept and related discourses through different disciplinary lenses can cater to a wider and more appropriate understanding of smart cities.

This book uses various cross-disciplinary approaches, introduced by Huang, Staupe, and Sageidet in chapter 3. The book's holistic and broad approach addresses the smart city as substance, content, and processes, including the 'hard' dimensions, such as information and communications technology (ICT) and infrastructure, as well as 'soft' dimensions such as citizen participation (cf. Wataya & Shaw, 2019). In relation to Stavanger, this holistic approach is also linked to the roadmap of the Smart Sustainable City Stavanger (Stavanger Municipality, 2016), as described in chapter 2 by Müller; chapter 1 by Aunemo, Undheim, and Sageidet; and chapter 10 by Undheim and Sageidet.

Smart and sustainable

There are essential interrelations between smart cities and sustainable development, defined by the Brundtland Report (World Commission on Environment and Development, 1987) and the Sustainable Development Goals (United Nations, 2015), acknowledged by the Roadmap for Norwegian Smart and Sustainable Cities (DOGA, Norwegian Smart City Network, & Nordic Edge, 2019), and currently promoted through the EU's climate-neutral and smart cities (Gaglione, 2023). Ahvenniemi et al. (2017) noted that there has been a shift from assessing urban performance according to sustainability indicators towards smart city goals, and Karvonen et al. (2020) stated that the smart city competes with and sometimes overshadows sustainable urban development agendas. Roggema (2020) revealed that 'smart' and 'sustainable' are used dichotomously and that research into these concepts seems divided. Therefore, multiple authors have called for a clearer integration or delineation between these concepts (Cugurullo, 2018; Karvonen et al., 2020; Roggema, 2020). The terms smart and sustainable are sometimes used synonymously or confusingly (Elgazzar & El-Gazzar, 2017). It is often unclear how the smart city concept relates to urban sustainable development or to what degree the smart city contributes to urban sustainability (Haarstad, 2017). It is therefore critical to understand how smart projects or strategies can impact urban life, urban behaviours, and urban societies at large (Khansari et al., 2014). It is also important to clarify and challenge the relationship between smart approaches and sustainable outcomes, something we are attempting to do in this book, as in chapter 4 by Lorentzen and Langhelle, in chapter 7 by Müller-Eie, and in chapter 11 by Nencioni, Sarah, and Müller.

Substance and process

Smart cities can be studied and implemented from different perspectives. Often, we look at how different elements can contribute to reaching a certain goal; for example, how a certain technology can contribute to improving the organizing of sustainable mobility or citizen safety (Winkowska et al., 2019). In other cases, the aim is to involve citizens in the development of a new neighbourhood to cater towards social inclusion and sustainability. In other situations, projects aim at developing or identifying the right process for a challenge, implying a more substantive understanding of processes. However, a more radical processual understanding (Cloutier & Langley, 2020) of smart cities may draw attention to what happens during and through the realization of different smart city activities and solutions, instead of focusing on what they consist of (Arin et al., 2023). In such a processual understanding, tangible and intangible elements have a place, and they are not seen as static input factors. They can change during the development processes, and their impact on the situation can also differ throughout.

In this book, we aim to show both substantial and processual approaches to the understanding of the smart city and its relationship with sustainable development, by taking a critical look at the underlying assumptions that are often taken for granted. Lindland, for instance, changes the linear and substantial understanding of the smart city in chapter 5, and Fisker, Müller, and Eiliott challenge assumptions about an imagined average in urban planning processes in chapter 6.

'Hard' and 'soft' domains

There seem to be distinct strands of smart city research. On the one hand, the origin of the smart city concept is grounded on technology (ICT) as a driver of urban efficiency and sustainability. This strand includes solutions with a techno-centric corporate smart city model, mainly applying to 'hard' domains such as buildings, energy grid, water, waste, and resource management, mobility, and logistics. However, there is a general lack of critical reflection on how urban development rooted in technological advancement and entrepreneurial thinking can impact the very nature of cities and urban life (Kitchin, 2015; Komninos et al., 2019). On the other hand, there is a call for a more holistic smart city view including soft domains such as policy, governance, culture, and education (Albino et al., 2015; Mora et al., 2017). Mora et al. (2017) claimed that part of the confusion or controversy around the existing heterogeneous smart city concepts originates from discrepancies among these different approaches and a resulting lack of common conceptualizations and definitions. Also, there is a lack of intellectual exchange between different fields and disciplines. Wiig (2015), for instance, claimed that the smart city can be used as a 'mask for entrepreneurial governance'

(p. 259), giving more focus to the capacity of a city to attract international corporations than a focus on local impact. Viewing cities as enterprises, they become driven by marked-oriented ideals of attractiveness rather than a locally grounded focus on sustainability and quality of urban environment and life. As an extension of that, Karvonen et al. (2020) advocated a need to recognize smart cities not as a technological agenda but rather as a socio-technical agenda that involves fundamental social, political, and cultural changes. Soft aspects of the smart city typically include a stronger focus on democracy, inclusion, justice, and livelihood – that is, all aspects of social sustainability – and with this, culture and art become elements in realizing more liveable and sustainable cities.

This book attempts to answer the call for different disciplines, fields, and research communities to integrate their efforts in smart city research to effectively counteract this currently growing divide (Mora et al., 2017) and includes both hard and soft aspects of smart cities, opening to a better understanding and stronger integration of these spheres, as exemplified in chapter 8 by Skotnes and Gould, in chapter 9 by Alberts and Fadnes, and in chapter 11 by Nencioni, Sarah, and Müller. Furthermore, chapter 14 by Sageidet, Kesarovski, and Zhivkov, addresses children as being and becoming smart and environmentally aware citizens.

Practice and theory

The places and spaces of smart cities can be used as testbeds or living labs for smart and sustainable solutions understood as pilots for exploring and demonstrating possible solutions in a realistic context (Cardullo et al., 2019). Therefore, much of the smart city literature is concerned with documenting and reporting on the testing of smart technology, applications, and processes. These living labs are not confined to physical and technical solutions; they also study how to realize better social conditions for, with, and by the citizens themselves, as discussed by Lorentzen and Langhelle in chapter 4, a detour from Stavanger to Trondheim. Methods for how to involve citizens in a democratic and just way and how to do this effectively are therefore emphasized (Paskaleva et al., 2015). Experiences have shown that it is often middle-class citizens of the ethnic and cultural majority who are engaged in co-creation community activities. This issue is challenged in chapter 6 by Fisker, Müller, and Eiliott.

Many of the previously mentioned aspects are, explicitly or implicitly, addressed by the roadmap of the Smart Sustainable City Stavanger (Stavanger Municipality, 2016) and by the international EU-funded research project Triangulum (Triangulum, 2019). Both the roadmap and the project Triangulum provide interesting starting points and examples for examining the unfolding of the smart city concept in practice and in a Nordic context. Triangulum and the Smart Sustainable City Stavanger have put the Stavanger region on the map as a resource centre for development in this regard. The Smart Sustainable City Stavanger underlines that curiosity for new knowledge, together with a variety of skills, is necessary to succeed in transforming creativity and promising ideas into specific innovations (Stavanger Municipality, 2016).

This book also provides an arena to share and critically reflect on experiences with analyzing, implementing, and assessing smart city approaches in a real Nordic urban context, as done in chapter 13 by Joudavi and Tarigan, in chapter 12 by Syse, Paneru, and Røstvik, and in chapter 15 by Lindland and Matre.

Structure and content of the book

This book represents an account of smart sustainable city approaches through a variety of disciplinary lenses, with some contributions describing and developing theoretical perspectives of the smart city within their field and others using smart city cases and applications to illustrate practice development. The book is therefore divided into three parts.

The first part of the book – A Smart City and a Smart Sustainable Cities Research Network – introduces how Stavanger became a smart city, as well as the Research Network for Smart Sustainable Cities at the UiS.

In chapter 1, Helga Aunemo (Stavanger Region European Office), Ståle Undheim (NIBIO), and Barbara Maria Sageidet (UiS) give an insight into how Stavanger, Norway, became one of the first smart cities with a research and practice agenda. Strategic decisions that led to Stavanger becoming a lighthouse city through the Triangulum project are highlighted. The chapter discusses how the smart city, in light of the intersection between European urbanization, sustainability urgency, and technological innovation, is an approach to deal with wicked urban challenges in a local context. Further, it suggests conducive conditions to integrating a smart city agenda with sustainable development, as well as to attracting EU research funding.

Chapter 2 is an essay, recording the inception and development of the Smart Sustainable Research Network at UiS, through the reflections of the network leader and coordinator Anders Riel Müller (UiS). This reflective narrative describes how targeted effort and financing can bring together researchers from different faculties and departments, structure collaboration between sectors, and create an open and responsive environment for knowledge exchange and development in an organization, as well as in the city and the region. Müller also highlights challenges that arise from working within different ontologies, epistemologies, methodologies, focuses, scopes, and timelines.

The second part of the book – The Conceptual Smart City – presents different perspectives on the theoretical and conceptual understanding of what the smart sustainable city is, could be, and should be through the lenses of sustainable transition and change management, social justice, urban planning, safety and security, and urban art.

In chapter 3, Stella Huang (LIU), Reidar Staupe (UiS), and Barbara Maria Sageidet (UiS) argue for the need for smart city research to be inherently cross-disciplinary. The chapter distinguishes different approaches of disciplinarity and explores research challenges and institutional barriers that inhibit research collaborations across disciplines and sectors. The chapter highlights the need for institutional structures and incentives in academic organization and funding bodies to support truly interdisciplinary smart city research.

In chapter 4 Veronika Lorentzen (UiS) and Oluf Langhelle (UiS) explore the smart city as a socio-technical transition with a side trip to Trondheim. They highlight the perspective of experimentation and innovation as a driver of societal change, sustainable development, and improvement. The chapter offers a view of how experimentation is a useful framework for enabling the emergence of ideas, practices, expectations, technologies, and social structures that induce change and contribute to sustainability.

Kristiane M.F. Lindland (NORCE/UiS) addresses ontological understandings and processes related to the smart city and points out how they predominantly follow a linear and substantive ontology. As an alternative understanding, chapter 5 presents a more relational and processual approach to the smart city concept, practice, and methods. This is underpinned through a theoretical foundation embedded in American Pragmatism, a re-interpretation of central themes of the smart city in that light, and a discussion of implications of a more processual understanding of the smart city as well as future research avenues.

Jens Kaae Fisker (UiS), Anders Riel Müller (UiS), and Helene Eiliott (Stavanger municipality) critically assess the notion that urban planning should target all citizens equally and therefore plan for an imagined average resembling themselves. They argue in chapter 6 that this average represents a minority of urban dwellers and instead contributes to sustaining a loop of inequality and exclusion that is detrimental to social sustainability. The chapter argues that if the smart city is to create better and more sustainable cities for all citizens, this mechanism of systematic inequality and exclusions needs to be overcome.

In chapter 7, Daniela Müller-Eie (UiS) gives insights into how the smart city can be viewed from an urban planning perspective, exemplified through concepts of sustainable urban mobility. In line with existing planning theory, whether the smart city concept should be viewed as a means or an end is discussed, with the conclusion that smart development must be a means to make cities more sustainable. The chapter also critically explores to which degree explorative and experimental smart approaches can be integrated with normative and deterministic planning theory.

Ruth Østgaard Skotnes (UiS) and Kenneth Pettersen Gould (UiS) address the management of risks in critical infrastructures as a crucial part of smart city development. In chapter 8, they discuss the organizational context that structures the relationship between the cyber system and the physical system of critical infrastructures in smart cities and regions. They argue that it is necessary to create multidisciplinary networks that overcome today's siloed organization and ensure a holistic approach to the safety and security management of these cyber-physical systems. The chapter discusses smart cities from an integrated safety and security perspective and contributes to a holistic and multidisciplinary view of the smart city concept.

In chapter 9, Kristoffer Berre Alberts (UiS) and Petter Frost Fadnes (UiS) take a rather radical and refreshing stand for what the smart city and smart art mean to the arts. They present a review of smart city terminology and what smart art is, and what it is not, by constituting a 10-point manifesto. Critically addressing smart – and dumb – in an artistic context and unpacking the risk of 'arts-washing', they also see a clear potential for smart art forcing artistic fields to connect and interact with all aspects of life as well as different kinds of local communities, seeing a possible empowerment of art through a smart lens. Although the use of smart art is both esoteric and controversial – especially within the art world itself – this chapter keeps the content grounded by drawing on a specific artistic project and its contribution to a smart art discourse.

The third part of the book – The Lived Smart City – uses concrete cases from the Smart Sustainable City Stavanger to exemplify what the smart city can be in a Nordic context, how it is established, what it means to be an inhabitant or researcher in a smart city, and which challenges may arise from that.

In chapter 10 Ståle Undheim (NIBIO) and Barbara Maria Sageidet (UiS) explore what happened with smart city ideas, as part of the Stavanger smart city agenda, during their implementation by the municipality's smart city office, in light of four theoretical perspectives on organizational change. The case study is supplemented with first-hand accounts of lived experiences with the follow-up of a smart city strategy from its inception to the field of practice. The study reveals challenges related to the practical implementation of smart city ideas and may inspire and critically prepare other medium-sized municipalities on their way to becoming a smart city.

In chapter 11, Gianfranco Nencioni (UiS), Annisa Sarah (UiS), and Anders Riel Müller (UiS) address wireless communication technology and the fifth generation (5G) of mobile networks and identify challenges related to the complexity in managing and orchestrating an integrated infrastructure, necessary for smart cities, based on experiences in Stavanger. Using artificial intelligence for data and resource orchestration in a UiS project, solutions are developed, investigated, and evaluated to overcome some challenges related to performance, flexibility, security, and dependability, while new challenges are posed.

While the smart city is often seen as a collector and provider of open-source data, Helleik Rosenvinge Syse (UiS), Chandra Prakash Paneru (UiS), and Harald Nils Røstvik (UiS) report in chapter 12 on the case of energy consumption data, investigating several barriers to data access. They find that there is a lack of solutions for simple anonymizing and for incentive structures for sharing of data, while data are often collected, managed, gate-kept, and monetized by private stakeholders. The situation is challenging, especially in relation to the research on energy generation, efficiency, and consumption. This is explored through a literature review on the availability of energy consumption data for researchers and a case study where the authors try to obtain access to energy data from various organizations.

In chapter 13 Ayda Joudavi (UiS) and Ari K.M. Tarigan (UiS) discuss the common barriers and motivators to cycling for daily commute and how smart solutions can be utilized to address these challenges of sustainable urban mobility. Additionally, the chapter explores the smart initiatives that have been adopted to promote cycling in Stavanger, Norway. Reviewing the adopted smart solutions, it investigates the barriers or motivators that have been targeted and those that have been left out. It summarizes by identifying existing knowledge gaps about the effectiveness of these smart solutions and suggests a future research agenda for how active travel can contribute to sustainable mobility futures. This chapter falls in the smart mobility domain and tries to enhance the broader comprehension of the smart sustainable city concept by depicting practical smart solutions to existing mobility challenges.

In chapter 14, Barbara Maria Sageidet (UiS), Todor Milkov Kesarovski (UiS), and Petar Zhivkov (Bulgarian Academy of Sciences) explore potentials in the Stavanger smart city to promote children's initial understandings of intertwined interrelations in their urban environment. To this end, they report on a survey among pedagogic leaders in some kindergartens in Stavanger about their awareness of air pollution as an environmental hazard as well as a teaching opportunity to introduce children to ecology, as well as technology, in their urban environment and enhance children's agency and beginning insights in the complexity of a smart sustainable city.

In chapter 15, Kristiane M.F. Lindland (NORCE/UiS) and Fredrik A.F. Matre (Public Policy Projects) give insights into how art and artistic methods are included in the smart sustainable city strategy in Stavanger and how this strategic aim is responded to in two different projects. Through insight from these projects, they highlight the potential challenges and possibilities of implementing more abstract visions into practical development work. Findings indicate that the smart art strategy can be used by stakeholders to reconsider traditional ways of doing things. Also, they may position themselves in new ways for novel collaborations. Following this, the chapter also gives insight into how the abstract idea of involving art and artistic methods in city development can be hampered by established norms and habits for what the contributions could and should be. As such, the chapter provides insight on how internalized social relations and norms can both inhibit and be re-interpreted through collaborations.

In the fourth part of the book – Lessons Learned – Chapter 16, by Daniela Müller-Eie (UiS), Barbara Maria Sageidet (UiS), and Kristiane M.F. Lindland (NORCE/UiS), summarizes the book by drawing on the different chapters to develop which lessons can be learned from the research of the Research Network for Smart Sustainable Cities at the UiS and experiences from the Stavanger smart city. Stavanger as a lighthouse city functions as a critical case that holds generalizable and transferable learning for others. The resulting 11 lessons highlight the need for inclusive cross-disciplinary and cross-sectoral collaboration and responsible and accountable use of smart sustainable approaches. The chapter provides a nuanced understanding of the complexities of translating policy into practice based on the examples of an existing smart city.

References

Ahvenniemi, H., Huovila, A., Pinto-Seppä, I., & Airaksinen, M. (2017). What are the differences between sustainable and smart cities? *Cities*, 60, 234–245. doi:10.1016/j.cities.2016.09.009

Albino, V., Berardi, U., & Dangelico, R.M. (2015). Smart cities: Definitions, dimensions, performance, and initiatives. *Journal of Urban Technology*, 22(1), 3–21. doi:10.1080/10630732.2014.942092

Arin, I.A., Supangkat, S.H., Gaol, F.L., & Ranti, B. (2023). Smart services implementation in smart cities: A comprehensive review of state-of-the-art technologies. In *Proceedings of the 2023 5th International Conference on Management Science and Industrial Engineering* (pp. 291–296). Association for Computing Machinery. doi:10.1145/3603955.3603971

Batty, M. (2017). *The age of the smart city*. Centre for Advanced Spatial Analysis (CASA) University College London. https://www.ucl.ac.uk/bartlett/casa/publications/2017/jun/casa-working-paper-210

Bibri, S.E. (2019). On the sustainability of smart and smarter cities in the era of big data: an interdisciplinary and transdisciplinary literature review. *Journal of Big Data*, 6, 25. doi:10.1186/s40537-019-0182-7

Bibri, S.E., & Krogstie, J. (2017). Smart sustainable cities of the future. An extensive interdisciplinary literature review. *Sustainable Cities and Societies*, 31, 183–212. doi:10.1016/j.scs.2017.02.016

Cardullo, P., De Feliciantonio, C., & Kitchin, R. (2019). *The right to the smart city*. Emerald. doi:10.1108/9781787691391

Cizelj, B. & Sinkovec, B. (2012). Knowledge cities and smart cities. *Knowledge Economy Network (KEN) Weekly Brief 20*, 1–4. https://pdf4pro.com/view/knowledge-cities-and-smart-cities-487563.html

Cloutier, C., & Langley, A. (2020). What makes a process theoretical contribution? *Organization Theory*, 1(1), 2631787720902473. doi:10.1177/2631787720902473

Cugurullo, F. (2018). Exposing smart cities and eco-cities: Frankenstein urbanism and the sustainability challenges of the experimental city. *Environment and Planning A: Economy and Space*, 50(1), 73–92. doi:10.1177/0308518X17738535

Dameri, R.P. (2013). Searching for smart city definition: A comprehensive proposal. *International Journal of Computers & Technology*, 11(5), 2544–2551. doi:10.24297/ijct.v11i5.1142

de Jong, M., Joss, S., Schraven, D., Zhan, C., & Weijnen, M. (2015). Sustainable–smart–resilient–low carbon–eco–knowledge cities; making sense of a multitude of concepts promoting sustainable urbanization. *Journal of Cleaner Production*, 109, 25–38. doi:10.1016/j.jclepro.2015.02.004

Derry, S., Schunn, C.D., & Gernsbacher, M.A. (Eds.). (2014). *Interdisciplinary collaboration: An emerging cognitive science*. Psychology Press. doi:10.4324/9781410613073

DOGA, Norwegian Smart City Network, & Nordic Edge. (2019). *Roadmap for smart and sustainable cities and communities in Norway. A guide for local and regional authorities developed by Design and Architecture Norway (DOGA), the Norwegian Smart City Network and Nordic Edge*. https://doga.no/globalassets/pdf/smartby-veikart-19x23cm-eng-v1_delt.pdf

Duygan, M., Fischer, M., Pärli, R., & Ingold, K. (2022). Where do smart cities grow? The spatial and socio-economic configurations of smart city development. *Sustainable Cities and Society*, 77, 103578. doi:10.1016/j.scs.2021.103578

Elgazzar, R.F., & El-Gazzar, R. (2017). Smart cities, sustainable cities, or both. In *Proceedings of the 6th International Conference on Smart Cities and Green ICT Systems* (pp. 250–257). SCITEPRESS. doi:10.5220/0006307302500257

Fairfield, J.D. (1994). The scientific management of urban space: Professional city planning and the legacy of progressive reform. *Journal of Urban History*, 20, 179–204. doi:10.1177/009614429402000202

Fernandez, T., Stöffler, S., & Diaz, C. (2023). 3-Triangulum: The three point project – Findings from one of the first EU smart city projects. In P. Droege (Ed.), *Intelligent environments* (2nd ed., pp. 87–107). Elsevier. doi:10.1016/B978-0-12-820247-0.00010-2

Gaglione, F. (2023). Policies and practices of transition towards climate-neutral and smart cities. *TeMA - Journal of Land Use, Mobility and Environment*, 16(1), 227–231. doi:10.6093/1970-9870/9822

Haarstad, H. (2017). Constructing the sustainable city: Examining the role of sustainability in the 'smart city' discourse. *Journal of Environmental Policy & Planning*, 19(4), 423–437. doi:10.1080/1523908X.2016.1245610

Karvonen, A., Cook, M., & Haarstad, H. (2020). Urban planning and the smart city: Projects, practices and politics. *Urban Planning*, 5(1), 65–68. doi:10.17645/up.v5i1.2936

Keshavarzi, G., Yildirim, Y., & Arefi, M. (2021). Does scale matter? An overview of the 'smart cities' literature. *Sustainable Cities and Society*, 74, 103151. doi:10.1016/j.scs.2021.103151

Khansari, N., Mostashari, A., & Mansouri, M. (2014). Impacting sustainable behavior and planning in smart city. *International Journal of Sustainable Land Use and Urban Planning*, 1(2), 46–61. doi:10.24102/IJSLUP.v1I2.365

Kitchin, R. (2015). Making sense of smart cities: Addressing present shortcomings. *Cambridge Journal of Regions, Economy and Society*, 8(1), 131–136. doi:10.1093/cjres/rsu027

Komninos, N., Kakderi, C., Panori, A., & Tsarchopoulos, P. (2019). Smart city planning from an evolutionary perspective. *Journal of Urban Technology*, 26(2), 3–20. doi:10.1080/10630732.2018.1485368

Menken, S., & Keestra, M. (Eds.). (2020). *An introduction to interdisciplinary research.* Amsterdam University Press. doi:11245/1.436989

Mora, L., Bolici, R., & Deakin, M. (2017). The first two decades of smart-city research: A bibliometric analysis. *Journal of Urban Technology*, 24(1), 3–27. doi:10.1080/10630732.2017.1285123

Paskaleva, K., Cooper, I., Linde, P., Peterson, B., & Götz, C. (2015). Stakeholder engagement in the smart city: Making living labs work. In M.P. Rodríguez-Bolívar (Ed.), *Transforming city governments for successful smart cities* (pp. 115–145). Springer. doi:10.1007/978-3-319-03167-5_7

Roggema, R. (2020). Towards integration of smart and sustainable cities. In R. Roggema & A. Roggema (Eds.), *Smart and sustainable cities and buildings* (pp. 5–23). Springer. doi:10.1007/978-3-030-37635-2

Sageidet, B.M., & Heggen, M.P. (2021). Global citizenship and the sustainable development goals. In W.L. Filho, P.G. Özuyar, A.M. Azul, L. Brandli, U. Azeiteiro, & T. Wall (Eds.), *Reduced inequalities. Encyclopedia of the UN Sustainable Development Goals* (pp. 1–11). Springer. doi:10.1007/978-3-319-71060-0_45-1

Shelton, T., Zook, M., & Wiig, A. (2015). The 'actually existing smart city'. *Cambridge Journal of Regions, Economy and Society*, 8(1), 13–25. doi:10.1093/cjres/rsu026

Späth, P., & Knieling, J. (2020). How EU-funded smart city experiments influence modes of planning for mobility: Observations from Hamburg. *Urban Transformations*, 2(1), 1–17. doi:10.1186/s42854-020-0006-2

Stavanger Municipality. (2016). *Roadmap for the smart city Stavanger – Vision, goals and priority areas.* roadmap-smart-city-stavanger-2016.pdf

Triangulum. (2019). *What is Triangulum?* https://triangulum.no/about-triangulum/?lang=en

UNHABITAT. (2022). *World cities report 2022. Envisaging the future of cities.* https://unhabitat.org/sites/default/files/2022/06/wcr_2022.pdf

United Nations. (2015). *Transforming our world: The 2030 agenda for sustainable development.* https://sdgs.un.org/2030agenda

Wataya, E., & Shaw, R. (2019). Measuring the value and the role of soft assets in smart city development. *Cities*, 94, 106–115. doi:10.1016/j.cities.2019.04.019

Wiig, A. (2015). IBM's smart city as techno-utopian policy mobility. *City*, 19(2–3), 258–273. doi:10.1080/13604813.2015.1016275

Winkowska, J., Szpilko, D., & Pejić, S. (2019). Smart city concept in the light of the literature review. *Engineering Management in Production and Services*, 11(2), 70–86. doi:10.2478/emj-2019-0012

World Commission on Environment and Development. (1987). *Our common future. A report from the United Nations world commission on environment and development.* Oxford University Press.

Zhao, F., Fashola, O.I., Olarewaju, T.I., & Onwumere, I. (2021). Smart city research: A holistic and state-of-the-art literature review. *Cities*, 119, 103406. doi:10.1016/j.cities.2021.103406

Part I

A smart city and a smart sustainable cities research network

1 From European sprat to European smart

How the fishing town of Stavanger became a smart city lighthouse

Helga Aunemo, Ståle Undheim and Barbara Maria Sageidet

Introduction

In this chapter, we seek to identify the enabling factors that laid the foundation for the smart city development of Stavanger, as part of the *smart city* trend unfolding in Europe. Within the European Smart Cities and Communities Project Triangulum – and along with the Dutch city of Eindhoven and the British city of Manchester – Stavanger claimed to fame as a smart city 'lighthouse' for Europe – and beyond (Padilla & Stöffler, 2017).

The chapter looks back on Stavanger's history as a rather small city in a European context. In view of European urbanization trends and related wicked problems (Okwechime et al., 2021; Zellner & Campbell, 2015), and based on practitioner experience and literature studies, this chapter explores Stavanger's journey towards becoming a smart city with the research question: How was the smart city concept implemented and realized in the City of Stavanger, and what were important factors in this process? Findings identified connections to Brussels and the European Commission *explore–shape–deal* framework as crucial to developing a proposal for Stavanger as a smart city. Further, business orientation, stakeholder relationships, networking, active citizen involvement, individuals as driving forces, and the integration of environmental perspectives into policy and innovation have been crucial factors to demonstrate the City of Stavanger's competence and capacities to contribute – as a lighthouse – to the further development of the smart city as an approach to dealing with wicked urban challenges in a local context. The added experiences may provide some lessons for future challenges in sustaining the smart city approach and achieving climate neutrality amidst disruptive economic and political changes.

Background and history

Development from a fishing town

Stavanger is today Norway's fourth largest city, with approximately 150,000 inhabitants (Stavanger Kommune, 2024). Its geographical position on the

DOI: 10.4324/9781003498650-3

Norwegian southwest coast, in the northern Jæren region, makes for a natural connection to the ocean, to the continent, and to the world. The city developed from a central farm settlement, traced back to Viking and early Mediaeval times, where a Bishop's seat was established from ca. AD 1120 (Brendalsmo & Paasche, 2017). Around 1800, Stavanger was still a fishing village of only 1,700 inhabitants, with its economy based on herring fisheries and later sea trade. This persisted until the early 20th century, when tinned herring from the canning industries gained international market traction (Jakobsen, 2017; Stangeland, 1984). By the 1960s, as the industrial sector was becoming more diversified, the town had 85,000 inhabitants. Stavanger was only moderately affluent, with citizens appearing 'provincial', and characterized by puritan religion and culture when vast resources of oil and gas were discovered in the 1970s (Stangeland, 1984). Stavanger became the national headquarters for the new offshore petroleum industry, entailing the establishment of major multinational oil companies and about 400 different service companies in the area (Øye Gjerde, 2002; Stangeland, 1984). In the following decades, the population continued to grow, with substantial international influx driven by the petroleum sector (Helle, 2016). In 1994, the Stavanger University College of Applied Sciences opened its doors, after a merger of several smaller schools, transitioning into the full-scale University of Stavanger by 2005. The establishment of academic and higher education institutions in the region has further developed Stavanger's ability to attract foreigners and to retain talent (Rørheim & Boschma, 2022). In 2008, the city celebrated the status as European Capital of Culture of the year, along with Liverpool, under the title 'an open port' (Bergsgard & Vassenden, 2011). In 2020, the City of Stavanger expanded its surface area, as it merged with neighbouring municipalities Finnøy and Rennesøy, as well as acquiring the remaining parts of the island Ombo. The metropolitan area of the greater Stavanger region includes the 12 neighbouring municipalities.

An insider's perspective

This study is a qualitative exploration of Stavanger's development towards becoming a smart city, using a participatory approach, broadly defined as bringing together various actors (van den Hove, 2000), underpinned with literature studies and a semi-structured interview, conducted in February 2022. The first author has been working in Brussels since 2016, most recently in the Stavanger Region European Office. The second author held a position as manager of the smart city department in the neighbour municipality, Sola. The informant, Heidi Kristina Jakobsen, was the director of the Stavanger Region European Office from 2012 until 2019, and the interviewer, Runa Monstad, held the same role from 2020 onwards. All of these participants have thus been partial insiders (Chavez, 2008), and boundaries between researchers, practitioners, and the researched are blurred, comparable to a qualitative insider approach (Greene, 2014).

The discussion gives relevance to the context and tries to critically elaborate reasoned reflections on complex issues; the authors are aware of the possible influence related to close physical, mental, and emotional relations to the study focus (Battersby & Bailin, 2011; Fook, 2011).

Contextual and theoretical frames

A shifting Europe: from rural to urban

More than 70% of the European population lives in urban areas (United Nations, 2019). An 'urban area' in this context is defined as a *city, town,* or *suburb* (Deuskar, 2015). In Europe, we see that disparities along the rural–urban divide are intensified as the population imbalance between remote and urban areas continues to grow (Tietjen & Jørgensen, 2016). Today, rural areas across Europe have lower education levels, fewer digital skills, and poorer health – partly explained by this rural-to-urban shift in demography (European Commission, 2021). Several measurements have been taken on the EU level to combat the challenges faced by Europe's rural populations, such as local innovation, job creation, lifelong learning, health initiatives, and attractive livelihoods. The EU Structural Funds, and later the EU Rural Action Plan and Pact, serve as examples (European Commission, 2021). Even if the attraction and resilience of urban areas are hard to beat (Giannakis & Bruggeman, 2020), a survey conducted in 2021 showed that young adults (aged 18–35) in the Nordic countries want to live in small cities (defined as urban places with less than 100,000 inhabitants; AFRY, 2021).

With the inclusion of the sustainability concept in both European and global development, cities with their population density and aggregated impact became obvious targets for climate action (European Commission, 2013). This becomes even more evident as we observe the young generation pointing at climate change as the biggest threat to future cities (AFRY, 2021).

European (smart) cities and participatory approaches

The *smart city* tradition in Europe is born into this trend of rising urban impact and sense of responsibility, coincident with the influx of technology, digitalization, and innovation (Angelidou, 2015). The role of technology and innovation in societal change is by no means new – they were drivers for the industrial revolution in Europe from 1750 onwards (More, 2002). The urbanization that followed, reinforced by increasing salaries and opportunities, led to new needs and demands from the urban population (Bairoch & Goertz, 1986).

The term *smart city* was first launched by Robert E. Hall in 1998, with a focus on information technology (Hall et al., 2000). Later, the concept was gradually expanded to include topics such as circular economy, infrastructure, architecture, construction, water and energy management, democracy, education, quality of life, culture, citizen involvement, management, and politics (Trivedi & Trivedi, 2024). This shows that the smart city concept builds on different disciplines and sheds light on collaboration between them (Albino et al., 2015; Calzada & Cobo, 2015).

A participatory approach is the *triple helix* concept, a model of innovation that uses the interactions between the three sectors of government, industry, and academia to foster social development and economic growth (Etzkowitz & Leydesdorff, 1995). The *quadruple helix* (Carayannis & Campbell, 2009) adds the perspective of the citizens to this approach.

The *smart city* concept later created a European momentum for *smart* urban development, to cope with common complex challenges. In recent decades, the sustainability issue has been added to this demand with the quest for healthy, liveable, and breathable urban spaces that provide optimized transport and smart solutions for its increasing number of inhabitants (European Commission, 2022). According to Haarstad et al. (2021), the Nordic cities are among the leading smart cities with regard to digitalization and stakeholder involvement.

Smart on the horizon: the explore–shape–deal framework

On their Smart City Marketplace Platform, a platform for smart city collaboration, the European Commission defines a smart city as 'a place where traditional networks and services are made more efficient with the use of digital solutions for the benefit of its inhabitants and business' (European Commission, 2022). The *network effect* is a phenomenon where the value a given participant derives from participating in the network increases with the number of participants (Jones, 2018).

The Smart City Marketplace saw the light of day when two former Commission smart city projects were merged: the Marketplace of the European Innovation Partnership on Smart Cities and Communities (EIP-SCC) and the "Smart Cities Information System" (SCIS). The new platform (as the two before) is a defining voice of the smart city concept and a go-to place to *explore* opportunities for smart city collaboration, to *shape* collaborative proposals, and to strike a deal; that is, an *explore–shape–deal* framework for matchmaking (European Commission, n.d.). This approach to smart city project development offers a process in which the knowledge transfer can flourish. The ultimate idea is for the know-how and experiences from capacity building, implementation, and replication to be optimized in a way that provides a model for future smart city projects.

During the Horizon 2020 framework period, there were two main sources of funding for such projects within the EU (Russo et al., 2014). Cities interested in promoting and further developing themselves as *smart* could either use private or public funding, such as the European Structural and Investment Funds (ESIF), or apply to become a lighthouse city through the Horizon 2020 programme directly. While ESIF is a major source of EU funding, channelling half of all money coming from the EU, this financial source is only open to EU member states. European cities from non-EU countries can, however, access EU programme funding if their country is *associated* with the programme in question. Through the EEA Agreement, Norway has been participating in the framework programmes for research and innovation since 1994, and Norwegian entities were able to participate in Horizon 2020 on the same terms as their EU counterparts.

The Triangulum project (Padilla & Stöffler, 2017), which saw the light of day in 2014, was a product of Horizon 2020 and the strategic focus on smart cities embedded within it. The idea was to establish so-called lighthouses for smart city implementation that could serve as testbeds and inspiration by sharing their best practice examples to their follower cities. As a flagship project, Triangulum was to put forward initiatives on sustainable mobility, energy, and information and communications technology under the slogan *demonstrate–disseminate–replicate.*

Wicked problems

The term *wicked problems*, defined by Rittel and Webber (1973), refers to major complex societal challenges for which it is impossible to find a single solution (Crowley & Head, 2017). It denotes a situation where elements are connected to different challenges that can be both large and extensive and solutions can only be expected through the use of methods other than the previously established often analytical, rational, and linear ways (Crowley & Head, 2017; Zellner & Campbell, 2015). Defining wicked problems is open to interpretation, as different parts of society have different perceptions of the importance of the complex challenges and how these can be solved (Lönngren & van Poeck, 2021). From a theoretical perspective, participatory processes have emerged as a necessary consequence of these complexities, especially on environmental issues (van den Hove, 2000). Head (2019) underlined the need for a broader approach, including creative thinking, problem framing, and public policy design, capacity, and implementation.

Implementation and realization of the smart city concept in the City of Stavanger and important factors

Explore: fish, oil – and EU funding?

The Norwegian capital of petroleum, Stavanger and its surroundings, is perceived by many as an international and outward looking region (Øye Gjerde, 2002; Rørheim & Boschma, 2022) and, in a European context, an early explorer of opportunities in the common market. It has been awarded with a substantial influx of both people and capital, over time creating a hub for know-how, investment, and innovation (Haarstad et al., 2021; Øye Gjerde, 2002). Such a position has required effort: to obtain, maintain, and remain an energy capital, the region has been required to identify opportunities, position itself, and *explore* (and exploit) the options at hand.

Stavanger has had a shift from canning industries to offshore petroleum industries (Stangeland, 1984). With the petroleum era set to gradually fade in the coming decades, the Stavanger region has begun the quest for alternative industries (Tripple et al., 2024), meeting (inter)national pressure for not only smarter but also more *sustainable* solutions (European Commission, 2013).

As Norway's membership in the European Union was up for debate for the second time in the early 1990s, the region quickly saw the importance of being present on the European stage. In 1993, the Stavanger Region European Office was established as the first Norwegian regional representation in Brussels. The national referendum the following year revealed a narrow majority vote for continuing European relations *outside* the framework of an EU membership (Bjørklund, 1997). Stavanger, however, continued its presence in Brussels, engaging in policy shaping and information flows, especially on matters regarding energy. According to Head (2019), such an anticipatory policy may be considered a creative and timely instrument to respond to (wicked) challenges. Complex challenges were exemplified by the City of Stavanger (2016, p. 3) as 'increasing urbanization, unsatisfactory infrastructure and inefficient mobility, increased competition for the best minds and the productive, profitable companies, greater demands and expectations from the inhabitants, environmental problems, higher technological change rate, and increasing proportion of older people in the population'. Such typical wicked problems (Rittel & Webber, 1973; cf. Tietjen & Jørgensen, 2016) are even more complex when including the greater Stavanger region (which is strongly developing from rural to urban), making Stavanger a highly relevant case for (smart) city projects and developments.

Collaboration and first-mover advantage

In Brussels, the Stavanger region joined a growing team of regions across the continent establishing themselves on the European arena (Huysseune & Jans, 2008). Leading up to 2014, the Stavanger Region European Office widened its scope into matters such as healthcare and health technology, digital solutions, and climate and environmental issues, reflecting political, technological, and academic trends and interests in Europe, as well as at home, and common efforts to solve intertwined urban challenges (Lönngren & van Poeck, 2021). Such a widened scope paved the way for new, strategic partnerships and alliances that were to prove valuable for future project collaboration (Russo et al., 2014).

Along with the launch of the European programme Horizon 2020 in 2014, which provided a record-breaking €80 billion in funding over 7 years, the value of these relationships, built over time, materialized in new ways. The City of Stavanger has long been enjoying good relations with its 11 *sister cities* around the world (Stavanger Kommune, 2024). With Horizon 2020's sine qua non of geographically spread sites and cross-border collaboration, the EU further paved the way for a more participatory in-depth cooperation (cf. van den Hove, 2000) to handle broadly wicked challenges of European character (cf. Rittel & Webber, 1973).

The idea was that the cross-border cooperation *process* (European Commission, n.d.), as much as the results, may lead to valuable insight that can be replicated in a regional, cross-sectoral context. Head (2019) considered such processes and insights as crucial learning dimensions to cope with wicked problems. Finding a common language, whether across geographical or sectoral divides, can lead to better understanding and, in turn, better learning outcomes that public administrations can use in the day-to-day communication with local stakeholders and communities (Heng et al., 2022).

No work but network!

Being present in Brussels allows organizations to take advantage of *network effects* (Jones, 2018). Brussels' added value vis-à-vis any given European capital is to a large extent the access to formal and informal interest groups and networks. And with networks comes information. Important Brussels-based networks within the (smart) city segment are Eurocities, the European Regions Research and Innovation Network (ERRIN), Cities and Regions for Transport Innovation (POLIS), and Covenant of Mayors, amongst others. These organizations are shaping urban and regional development, providing platforms on which European regions can engage with each other across the continent. Through these platforms, members are encouraged to share views and experiences on both policy and project development through various activities and to find potential partners who have similar challenges to those their regions may be facing (Trobbiani, 2015).

With the introduction of Horizon 2020, the problem-solving aspect of research and innovation became more central than in previous EU framework programmes. Research and innovation (R&I) were to serve the greater European society, and with the increasing investments in R&I at the European level, expectation of tangible results followed. Concrete plans as to how the results could lead to better solutions benefitting society at large, an issue highlighted by Head (2019), became a prerequisite for funding. This in turn came to determine how project facilitators were approaching new opportunities. 'Walking across the aisle' and ensuring broad involvement became a prerequisite for successful proposals (cf. van den Hove, 2000).

Shape: creating the future

The region of Stavanger has been considered one with a strong work ethic, with close ties between the various stakeholders (City of Stavanger, 2016). Its relatively small and tight-knit organizational shape makes for a region in which actors in various fields know each other and make alliances across public–private divisions. Together they *shape* the solutions for the future and collaborate on finding the best answers to the challenges at hand. This is in line with more recent understandings of fruitful responses to wicked problems (Crowley & Head, 2017; Head, 2019).

Adding a strong academic foundation to the mix, led by the University of Stavanger and joined by a wide range of sectoral clusters and technological know-how, the region has a strong *triple helix* base (Etzkowitz & Leydesdorff, 1995) for cross-cutting collaboration, perhaps not unknown in the Nordic countries – home to *three-party cooperation* and *the Nordic model* (Kangas & Kvist, 2019; SAMAK, 2016). At a regional level, however, such tightly knit configurations were rather unique: whereas Norwegian regions in general had a strong public and academic sector influence on their own development, the Stavanger region has for a long time had an equally strong business orientation (H.K. Jakobsen, personal communication, February 18, 2022). This became evident when the region came together to position itself at the national level for the title of 'oil capital' in the late 1960s and early 1970s (Øye Gjerde, 2002). Adjusting to the industry's needs, through dialogue and collaboration, became a prerequisite for success. Another more recent example of the business-oriented approach in Stavanger is the regional hydropower company Lyse AS, which, in light of its inter-municipality ownership, has diversified its business into telecom, becoming a fibre optics internet provider at the national scale (Zachariassen, 2015). This provided Lyse AS with the necessary technological capacity to strengthen Stavanger's Horizon 2020 smart city proposal.

A people-centred approach to urban development

A sudden shift away from brisling (European sprat) to a petroleum-driven economy in the early 1970s boosted the local demand side, creating an influx of skills, investments, and innovation, crucial for the further growth in the sector. The skills and capital, in the form of labour and cash investments, has developed a fourth dimension – the citizens – to be added to the cross-sectoral cooperation in Stavanger. Starting with early labour immigration from historically petroleum-savvy countries such as the United States and France (Roalkvam & Øye Gjerde, 2012), the new Norwegian oil economy soon attracted industry and knowledge workers alike. Partly as a result, Stavanger is today an intercultural city, with people from various nationalities, origins, languages, and religious orientations. In 2014 Stavanger was fourth among 60 European cities on the Intercultural Cities Index, which measures the degree of interculturality, with an aggregate score of 83% (Council of Europe, 2014). Strategic efforts at the municipal level, encouraging interaction between diverse groups in public spaces, was part of the justification, pointing to the fact that more than 15% of the inhabitants in Stavanger are foreign born. UNESCO (2005, p. 14) defines *interculturality* as 'the existence and equitable interaction of diverse cultures and the possibility of generating shared cultural expressions through dialogue and mutual respect'. However, there are also challenges related to interculturality in Stavanger (see Fisker et al., chapter 6).

Underlining *interaction*, the idea of cross-sector collaboration as seen between the public sector, industry, and academia is increasingly extended to include the wider public, or the citizens, as a fourth interest group (Cai & Lattu, 2021). The City of Stavanger has been a national driving force through the practical use of *triple* and *quadruple helices* (Carayannis & Campbell, 2009; Etzkowitz & Leydesdorff, 1995) in developing its EU smart city proposal, as well as improving local citizen involvement and decision making. The use of citizen panels across various demographic groups is one such example. The quadruple helix approach not only acknowledges the citizens as an essential stakeholder group but defines it as one of equal standing with academia, the public sector, and industry, pointing out the need for the three latter to *interact* with the citizens on equal footing. People, being the end users of any public services and innovative solutions spinning out from a smart city project, may provide early input in a consultation process, helping to identify obstacles that might otherwise have been overlooked (see also Fisker et al., chapter 6).

Citizens' (community) involvement and the quadruple helix approach to project development and implementations have gained traction in Brussels, gradually increasing in importance with Horizon Europe, the 2021 and onwards successor of Horizon 2020. The interculturality, along with the regional drive for continuous innovation and exploration, may help explain why Stavanger, and the region as a whole, was well positioned to meet the opportunities in the EU framework programmes in general and the Triangulum project in particular. The triple (and quadruple) helix approach is further exemplified by the diverse membership base engaging in the Brussels office since the beginning in 1993. Such a wide, regional gathering of interests in Brussels simultaneously reinforced alliances back home. These alliances were in many ways crucial when the Triangulum opportunity arose.

Seal the deal: driving forces

From exploring and shaping, any project concept needs to culminate in a deal to get off the drawing table. The director of the Brussels office at that time, Heidi Kristina Jakobsen, illustrates how a single person with the right drive and network can become a key player in international project development. After having *explored* and *shaped* the project opportunities in Triangulum along with regional and European stakeholders, she was now ready to *strike* a deal fit for Stavanger. Engaging with stakeholders back home, emphasizing the opportunities embedded in a lighthouse project like Triangulum, she could capitalize on a vast European network, thanks to her time at Eurocities, amongst others (H.K. Jakobsen, personal communication, February 18, 2022). At this point the office had been present in Brussels for 20 years, with access to the relevant circles, early information, and potential project partners. An important aspect of this process was to ensure political backing by engaging key players on the political arena. As the Brussels office at the time was liaised with Greater Stavanger, an earlier regional business development hub, the support and follow-up from key people who saw the potential was key. Tone Grindland, in Greater Stavanger at that time, who had herself experience from the regional Brussels office and kept up the pressure at home, was invaluable (H.K. Jakobsen, personal communication, February 18, 2022). Further, it required political courage locally to try something new and to be willing to open up to a European opportunity such as Triangulum.

In 2014, the City of Stavanger, in collaboration with Lyse AS, the Rogaland County Council, the University of Stavanger, and Greater Stavanger, was awarded Horizon 2020 funding, together with the cities of Manchester (Great Britain) and Eindhoven (The Netherlands).

Such strong and dynamic links between Brussels and the EU agenda on one side and the political and administrative community in the region on the other might have been a winning component in creating the regional consensus needed to strike a deal in Europe. Similarly, the office, through its long-time presence in Brussels, had the network needed to discover and explore the opportunity and become part of a strong consortium. The decades-long presence, positioning, and people orientation of the Stavanger Region European Office had borne fruits.

Smart ... and sustainable?

The Global Sustainable Development Goals (United Nations, 2015) require a shift towards alternative sources of energy that will equally supply jobs and demand local know-how. To promote this transition, smart and strategic thinking, as applied in Triangulum, is needed. Having declared the 2020s 'EU's Digital Decade', the European Commission is approaching the challenges ahead with policy and regulation aimed at incentivizing active, yet safe use of data and technology. In a smart city context, the use of AI, the collection (and use) of user data, and cyber security are crucial aspects to be addressed (see Nencioni et al., chapter 11, and Skotnes and Gould, chapter 8). Applying collective data in decision-making processes in urban planning allows for targeted, user-oriented decisions that in turn can lead to resilient, sustainable, and inclusive solutions and services. Making use of data available can be an effective way of addressing societal issues, in addition to more traditional forms of citizen involvement.

One such societal issue is Nordic young adults' change in perceptions when it comes to housing and living and their preference for small cities (AFRY, 2021). This shift away from the megacity idea towards smaller urban centres is explained by shifting perceptions of lifestyle ideals (Metzger et al., 2018). Changes in work patterns and the re-emergence of the *bikeable* and *15-minute* city concepts (see Joudavi and Tarigan, chapter 13) might have led to new personal priorities amongst the young. Having seen the day of light with the green and digital transitions (Metzger et al., 2018), those trends were accelerated in unforeseen ways by the pandemic (Francke, 2022). According to Metzger et al. (2018), more young people foresee the opportunity to have a garden space. For a city like Stavanger and other European cities alike, this could mean possible future trade-offs between local and sustainable food production and less centralized housing policies, unless these policies are thoughtfully integrated.

Smart with a heart

Smart city initiatives have significant impacts on the lives of citizens (McKinsey Global Institute, 2018). A human framework, acknowledging the importance of creativity, learning, humanity, and knowledge in city development, has gained traction in Europe (Evans et al., 2019; cf. Head, 2019). The Nordic Edge Expo tagline from 2018, *Smart With a Heart*, illustrates the trend. The New European Bauhaus initiative is another, more recent example (see Fisker et al., chapter 6). Launched in 2021 and inspired by the original Bauhaus movement emerging in the interwar period, it has spun out from the Green Deal, emphasizing the importance of *beautiful, sustainable,* and *inclusive* livelihoods. In the future, all policy, innovation, and research initiatives in the EU (and in Europe) will have to take both human and environmental considerations into account, rather than trading the two off against more short-term financial gains. The strong emphasis on *citizen involvement, gender,* and *resilience* seen in Horizon Europe reflects the political discussion and societal trends at large. Such trends have equally influenced the smart city paradigm (Albino et al., 2015; Calzada & Cobo, 2015).

With Horizon Europe, the idea of *missions*, as conceptualized by economist Marianna Mazzucato (2021), has gained traction in Europe. Missions are a set of societal targets or common goals, aiming at tackling some of the most pressing challenges of our time (United Nations, 2015; cf. Lönngren & van Poeck, 2021). The Commission, applying this targeted, problem-solving logic in Horizon Europe, identified five such missions to be addressed in the coming years. One of them, *Mission Cities*, aims for 112 of Europe's cities to become climate-neutral by 2030, and Stavanger is now one of them (European Commission, 2024).

The sustainability aspect has thus become increasingly visible in the smart city discourse, as we have moved from technology optimism in Horizon 2020 to a greener and climate-neutral perspective as we look towards 2030, underpinning the idea that without a healthy climate, humanity cannot thrive.

Conclusion

Through the Triangulum project, the City of Stavanger has tried to *demonstrate* its role as European lighthouse for urban planning and development and may *disseminate replicable* lessons to further develop an understanding and suitable measures to address the wicked social and environmental challenges in European medium-sized municipalities.

In Stavanger, the *smart city* concept was developed and realized mainly through the region's early connections to Brussels and an international network (early *exploration*), through the *shaping* of crucial stakeholders and project ideas, and, later, submitting a project application – *the deal*. Further, the role of networking and collaboration is exemplified by the triple and quadruple helix approaches.

Important factors for this process were Stavanger's business orientation, stakeholder relationship, intercultural qualities, individual driving forces,- and the pivotal role of citizen involvement. People are fundamental to successful smart city development – in Stavanger and beyond. Without the citizens in the driver seat, inclusive, long-lasting results might be hard to come by.

The integration of human and environmental perspectives into policy, technological development, and innovation has been revealed as crucial to promote inclusivity and the green transition and Stavanger's goals towards climate neutrality.

With the people at the core of the process, Europe can create cities that are smart and resilient, where people thrive socially and mentally, in healthy, beautiful neighbourhoods. Or in the words of Bertholdt Brecht (1984, p. 1010): 'But what are cities, built without the wisdom of the people?'

References

AFRY. (2021). *Future cities: Survey 2021.*
Albino, V., Berardi, U., & Dangelico, R.M. (2015). Smart cities: Definitions, dimensions, performance, and initiatives. *Journal of Urban Technology*, 22(1), 3–21. doi:10.1080/10630732.2014.942092
Angelidou, M. (2015). Smart cities: A conjuncture of four forces. *Cities*, 47, 95–106.

Bairoch, P., & Goertz, G. (1986). Factors of urbanisation in the nineteenth century developed countries: A descriptive and econometric analysis. *Urban Studies*, 23(4), 285–305.

Battersby, M., & Bailin, S. (2011). Critical inquiry: Considering the context. *Argumentation*, 25, 243–253. doi:10.1007/s10503-011-9205-z

Bergsgard, N., & Vassenden, A. (2011). The legacy of Stavanger as Capital of Culture in Europe 2008: Watershed or puff of wind? *International Journal of Cultural Policy*, 17(3), 301–320.

Bjørklund, T. (1997). Old and new patterns: The 'no' majority in the 1972 and 1994 EC/EU referendums in Norway. *Acta Sociologica*, 40(2), 143–159. http://www.jstor.org/stable/4201020

Brecht, B. (1984). Big time lost. In N. Venus & B. Reynolds (Eds. & Trans.), *Die Gedichte* (p. 1010). Suhrkamp.

Brendalsmo, J., & Paasche, K. (2017). Stavanger – Før det ble en by [Stavanger – Before it became a city]. *Historisk Tidsskrift*, 96(2), 103–123.

Cai, Y., & Lattu, A. (2021). Triple helix or quadruple helix: Which model of innovation to choose for empirical studies? *Minerva*, 60, 257–280. https://link.springer.com/article/10.1007/s11024-021-09453-6

Calzada, I., & Coba, C. (2015). Unplugging: Deconstructing the smart city. *Journal of Urban Technology*, 22(1), 23–43. doi:10.1080/10630732.2014.971535

Carayannis, E.G., & Campbell, D.F.J. (2009). 'Mode 3' and 'quadruple helix': Towards a 21st century fractal innovation ecosystem. *International Journal of Technology Management*, 46(3–4), 201–234.

Chavez, C. (2008). Conceptualizing from the inside: Advantages, complications, and demands on insider positionality. *The Qualitative Report*, 13(3), 474–494. http://www.nova.edu/ssss/QR/QR13-3/chavez.pdf

City of Stavanger. (2016). *Roadmap for the smart city Stavanger – Vision, goals and priority areas.* roadmap-smart-city-stavanger-2016.pdf

Council of Europe. (2014). *Stavanger: Results of the Intercultural Cities Index.*

Crowley, K., & Head, B.W. (2017). The enduring challenge of wicked problems: Revisiting Rittel and Webber. *Policy Science*, 2017(50), 539–547. doi:10.1007/s11077-017-9302-4

Deuskar, C. (2015, June 2). *What does 'urban' mean?* [Blog entry]. World Bank Blogs. https://blogs.worldbank.org/sustainablecities/what-does-urban-mean

Etzkowitz, H., & Leydesdorff, L. (1995). The triple helix – University–industry–government relations: A laboratory for knowledge based economic development. *EASST Review*, 14, 14–19.

European Commission. (n.d.). *Smart Cities Marketplace.* https://smart-cities-marketplace.ec.europa.eu/

European Commission. (2013). *The EU strategy on adaptation to climate change.* https://ec.europa.eu/clima/system/files/2016-11/eu_strategy_en.pdf

European Commission. (2021). *Communication from the Commission to the European Parliament, the Council, the European Economic and Social Committee and the Committee of the Regions: A long-term vision for the EU's rural areas – Towards stronger, connected, resilient and prosperous rural areas by 2040.*

European Commission. (2022). *Smart cities.*

European Commission. (2024). *EU missions, 100 climate-neutral and smart cities – Cities on a journey to climate neutrality.* https://data.europe.eu/doi/10.2777/169604

Evans, J., Karvonen, A., Luque-Ayala, A., Martin, C., McCormick, K., Raven, R., & Palgan, Y.V. (Eds.). (2019). Smart and sustainable cities? Pipedreams, practicalities and possibilities. *Local Environment*, 24(7), 557–564.

Fook, J. (2011). Developing critical reflection as a research method. In J. Higgs, A. Titchen, D. Horsefall, & D. Bridges (Eds.), *Creative spaces for qualitative researching* (pp. 55–65). Sense.

Francke, A. (2022). Cycling during and after the COVID-19 pandemic. *Advances in Transport Policy and Planning*, 10, 265–290.

Giannakis, E., & Bruggeman, A. (2020). Regional disparities in economic resilience in the European Union across the urban–rural divide. *Regional Studies*, 54(9), 1200–1213, doi:10.1080/00343404.2019.1698720

Greene, M.J. (2014). On the inside looking in: Methodological insights and challenges in conducting qualitative insider research. *The Qualitative Report*, 19(15), 1–13. http:// www.nova.edu/ssss/QR/QR19/greene15.pdf

Haarstad, H., Hansen, G.S., Andersen, B., Harboe, L., Ljunggren, L., Røe, P.G., Wanvik, T. I., & Wulff-Wathne, M. (2021). Nordic responses to urban challenges of the 21st century. *Nordic Journal of Urban Studies*, 1(1), 4–18. doi:10.18261/issn.2703-8866-2021-01-0

Hall, R.E., Bowerman, B., Braverman, J., Taylor, J., Todosow, H., & van Wimmersberg, U. (2000, September 28). *The vision of a smart city* [paper presentation]. Second International Life Extension Technology Workshop, Paris.

Head, B.W. (2019). Forty years of wicked problem literature: Forcing closer links to policy studies. *Policy and Society*, 38(2), 80–97. doi:10.1080/14494035.2018.1488797

Helle, K. (2016). *The history of Stavanger: A complete urban history*. Wigestrand.

Heng, S., Cheng, D., Tsilionis, K., & Wautelet, Y. (2022). Stakeholder-based management of smart cities: The case of Brussels. In A. Visyizi & O. Troisi (Eds.), *Managing smart cities* (pp. 213–229). Springer. doi:10.1007/978-3-030-93585-6_12

Huysseune, M., & Jans, M. (2008). Brussels as the capital of a Europe of the regions? Regional offices as European policy actors. *Brussels Studies*, 16, 1–11.

Jakobsen, S.E. (2017, September 8). *Da Stavanger var en hermetikkby*. Norges Forskningsråd. https://forskning.no/historie-norges-forskningsrad-partner/da-stavanger-va r-en-hermetikkby/324424

Jones, G. (2018). *The new Palgrave dictionary of economics*. Palgrave Macmillan.

Kangas, O., & Kvist, J. (2019). Nordic welfare states. In B. Greve (Ed.). *Routledge handbook of the welfare state* (2nd ed.). Routledge.

Lönngren, J., & van Poeck, K. (2021). Wicked problems: A mapping review of the literature. *International Journal of Sustainable Development and World Ecology*, 28(6), 481–502. doi:10.1080/13504509.2020.1859415

Mazzucato, M. (2021). *Mission economy: A moonshot guide to changing capitalism*. Allen Lane.

McKinsey Global Institute. (2018). *Smart city technology for a more liveable future*. http s://www.mckinsey.com/business-functions/operations/our-insights/smart-cities-digita l-solutions-for-a-more-livable-future

Metzger, M.J., Murray-Rust, D., Houtkamp, J., Paterson, J.S., Valluri-Nitsch, C., La Riviere, I., Pérez-Soba, M., & Jensen, A. (2018). How do Europeans want to live in 2040? Citizen visions and their consequences for European land use. *Regional Environmental Change*, 18, 789–802.

More, C. (2002). *Understanding the industrial revolution*. Routledge.

Okwechime, E., Duncan, P.B., Edgar, D.A., Magnaghi, E., & Vegliarti, E. (2021). Smart cities: A response to wicked problems. In E. Magnaghi, V. Flambard, D. Mancini, J. Jacques, & N. Gouvy (Eds.), *Organizing smart buildings and cities* (pp. 13–34). Springer.

Øye Gjerde, K. (2002). *Stavanger er stedet: Oljeby 1972–2002*. Norsk Oljemuseum.

Padilla, M., & Stöffler, S. (2017). On the path towards smart mobility: The journey of three forerunner cities Eindhoven, Manchester and Stavanger. In M. Schrenk, V.V.

Popovich, P. Zeile, P. Elisei, & C. Beyer (Eds.), *REAL CORP 2017-Panta-Rhei-A world in constant motion. Proceedings of the 22nd International Conference on Urban Planning. Regional Development and Information Society* (pp. 383–390).

Rittel, H.J., & Webber, M. (1973). Dilemmas in a general theory of planning. *Policy Sciences*, 4(2), 155–169. doi:10.1007/BF01405730

Roalkvam, G., & Øye Gjerde, K. (2012). *Stavanger bys historie: Oljebyen 1965–2010: Vol. 4: The oil city 1965–2010* (H. Hamre & K. Helle, Eds.). Wigestrand.

Rørheim, J.-E., & Boschma, R. (2022). Skill-relatedness and employment growth of firms in times of prosperity and crisis in an oil-dependent region. *Economy and Space*, 54(4), 676–692. doi:10.1177/0308518X211066102

Russo, F., Rindone, C., & Panuccio, P. (2014). The process of smart city definition at an EU level. *WIT Transactions on Ecology and the Environment*, 191, 979–989.

SAMAK. (2016). *The Nordic model for dummies: All you need to know in 6 minutes.* http s://www.finansforbundet.no/content/uploads/2020/10/THE_NORDIC_MODEL_FOR_DUMMIES.pdf

Stangeland, P. (1984). Getting rich slowly – The social impact of oil activities. *Acta Sociologica*, 27(3), 215–237. doi:10.1177/000169938402700304

Stavanger Kommune. (2024). *Fakta om Stavanger.* https://www.stavanger.kommune.no/om-stavanger-kommune/fakta-om-stavanger/#stavangers-vennskapsbyer

Tietjen, A., & Jørgensen, G. (2016). Translating a wicked problem: A strategic planning approach to rural shrinkage in Denmark. *Landscape and Urban Planning*, 154, 29–43. doi:10.1016/j.landurbplan.2016.01.009

Tripple, M., Fastenrath, S., & Isaksen, A. (2024). Rethinking regional economic resilience: Preconditions and processes shaping transformative resilience. *European Urban and Regional Studies*, 31(2), 101–115. doi:101177/09697764231172326

Trivedi, A., & Trivedi, N. (2024). Integrating circular economy in smart cities: Challenges and pathways to sustainable urban development. In V. Kandpal, E.D. Santibanez-Gonzales, P. Chatterjee, & M.K. Nallapaneni (Eds.), *Smart cities and circular economy* (pp. 71–82). Emerald. doi:10.1108/978-1-83797-957-820241006

Trobbiani, R. (2016). *European regions in Brussels: Towards functional interest representation?* [Bruges Political Research Papers 53/2016]. Department of European Political and Administrative Studies.

UNESCO. (2005). *Convention on the protection and promotion of the diversity of cultural expressions.*

United Nations. (2015). *Transforming our world: The 2030 agenda for sustainable development.* https://sdgs.un.org/2030agenda

United Nations. (2019). *World urbanization prospects: The 2018 revision.* doi:10.18356/b9e995fe-en

van den Hove, S. (2000). Participatory approaches to environmental policy-making: The European Commission climate policy process as a case study. *Ecological Economics*, 33, 457–472. doi:10.1016/S0921-8009(99)00165-2

Zachariassen, E. (2015). *Hadde vi visst hvor stort Altibox ville bli, hadde vi ikke våget å starte.* Digi.no. https://www.digi.no/artikler/hadde-vi-visst-hvor-stort-altibox-ville-bli-hadde-vi-ikke-vaget-a-starte/209008

Zellner, M., & Campbell, S.D. (2015). Planning for deep-rooted problems: What can we learn from aligning complex systems and wicked problems. *Planning Theory & Practice*, 16(4), 457–478. doi:10.1080/14649357.2015.1084360

2 Smartening Stavanger – reflections on interdisciplinary and intra-regional collaboration

Anders Riel Müller

I came to Stavanger in 2019 to lead the Research Network for Smart and Sustainable Cities. Coming from outside of Stavanger and even outside of Norway with little knowledge of both Norway and especially the city of Stavanger has both its advantages and challenges. Getting to learn about the city and the region, its history, economy, politics, and not least its smart city initiatives has been an enlightening experience. This chapter provides an overview of the background for the focus on smart city policies, strategies, and activities in Stavanger and reflections on how the smart city agenda has shaped research activities and collaboration with the public and private sectors in the region. These reflections are centred on the background for the emergence of *smart city* as a regional development strategy and how the fuzziness of the smart city concept has provided new spaces for collaboration.

Stavanger became one of the first cities to receive funding from the EU's Smart Cities Lighthouse program back in late 2014. The Triangulum project received 25 million euros to develop testbeds for innovative projects focusing on sustainable mobility, energy, information and communications technology (ICT), and business opportunities in Stavanger, Eindhoven, and Manchester[1] (Triangulum, 2014). The project, which ran from 2015 to 2020, was the largest EU project the city had ever been part of. Triangulum also signalled a new potential direction for economic development and to rebrand the city as more than an oil city in the eyes of the world. Triangulum was not the first attempt at this. In 2008, Stavanger was European Capital of Culture, which can be interpreted as an attempt to elevate the city's standing as more than a city of conspicuous materialist consumption and new wealth but also as a regional economic development strategy to diversify the economic base towards the creative and cultural sectors (Asle Bergsgard & Vassenden, 2011; Marco-Serrano et al., 2014). The ambition to become a home of creative industries was concurrent with another initiative to rebrand Stavanger from the oil capital to the energy capital (Nygaard, 2009).

The awarding of the Smart Cities Lighthouse project coincided with the drastic drop in global oil prices in 2015. From 2014 to 2015, Brent crude oil prices dropped almost 50%. Stavanger experienced a severe economic downturn as a result. In August 2015, it was estimated that more than 12,000 jobs would be lost in the region of Stavanger alone, and this was only in sectors directly involved in

DOI: 10.4324/9781003498650-4

oil and gas activities (Kongsnes, 2015). Other sectors were also being impacted, most notably service industries such as the hotel and restaurant sectors. Stavanger's high reliance on oil and gas as its economic base had been a concern for decades, but the drop in oil prices in 2015 was a blow to the regional economy that dwarfed the previous major economic crisis of the late 1980s (Nerheim, 2014).

It is with this economic crisis as the backdrop that the attention to *smart city* as an economic development opportunity should be understood. With the awarding of the Smart Cities Lighthouse project, Stavanger spotted an opportunity to diversify its economic base through becoming an innovation hub for smart city solutions (Munkvik, 2014; Øvrebekk, 2017). In 2015, Nordic Edge Expo had been launched with the ambition to become the largest smart city expo and conference in Northern Europe (Stavanger Aftenblad, 2016), following in the footsteps of the Barcelona Smart City Expo World Congress, which started in 2011. The city followed up by establishing its own smart city office in 2017 (see Undheim and Sageidet, chapter 10) to lead municipal smart city initiatives across different sectors and to strengthen collaboration regionally, nationally, and internationally.

The University of Stavanger was a relative latecomer to the local smart city arena. Although some researchers were involved in the Triangulum project, there was no significant research activity on smart cities outside the Triangulum project. With increased pressure from regional industry and the public sector, the university launched a series of workshops in 2018 to gauge interest amongst researchers in making a strategic initiative on smart cities. The workshops garnered significant attention from different research environments, and it was decided to establish an institution-wide research network that would encourage close collaboration with regional stakeholders and encourage interdisciplinary collaboration across faculties and departments. Furthermore, to kick-start the research network, seven PhD scholarships were announced, distributed across faculties.

Working across disciplines

A core mandate for the newly established network was to encourage interdisciplinary collaboration among researchers at the university on smart city topics. The organizational structure of the university is far from conducive to such collaboration, and many researchers are not familiar with research outside their own research group and department, much less at other faculties. Developing a research environment that cuts across the organizational silo structure is the art of creating a matrix structure with no other motivation than individual interest. Researchers are not allocated time to participate in network activities. This is a tough sell, as most researchers already are busy with teaching and research and caught up in other department-level or faculty-level initiatives to which they have been allocated time.

Furthermore, bringing together research from very different fields with different ontological, epistemological, and methodological approaches and intellectual traditions can easily lead to misunderstandings. For example, while some see it as important to define what a smart city is, others are more interested in understanding what the smart city does. Bringing together researchers based in positivist science with people ascribing to more constructivist approaches requires openness and generosity; nevertheless, it may be hard to find common ground. The other challenge is that research on *smart city* can be either very narrow or extremely broad. While smart cities in their inception mostly focused on the deployment of ICT to address urban problems, the smart city concept has come to mean a whole host of other things. For example, in the roadmap of the Stavanger Smart City, smart city is more of an approach to solving urban problems that may include ICT but first and foremost emphasizes 'smart' solutions in which *smartness* is defined as a process and mode of governing to better address wicked urban problems. Smartness in the roadmap is defined as a combination of technology, collaboration, and citizen involvement while covering a whole range of topics including education, health, culture, energy, climate, environment, and democracy (Stavanger Municipality, 2016).

The research network decided on a similarly broad approach to smart city research. There were several reasons for this. Firstly, there was a desire to engage a broad segment of researchers and, secondly, there was a wish to align research themes with the strategic priorities of the municipality and Nordic Edge, as this book also shows. This led to a (very) broad scope for network activities. In the end, eight topics were identified based on interest among the different researchers and the interests of regional partners:

- Education and knowledge
- Arts and culture
- Climate and sustainability
- Mobility and transport
- Safety
- Democracy and participation
- Innovation and entrepreneurship
- Data and telecommunications

These topics are also evident in the contributions to this book. In practice, most research activities have cut across two or more of these thematic topics, and it has been the cross-cutting engagements that have been most interesting to observe. Encounters between different fields of expertise and topical interests have been needed in many collaboration projects with regional partners, as urban and regional challenges are rarely possible to address through a single discipline or approach (see Huang et al., chapter 3). This is not without challenges, but movement only occurs through friction. So, rubbing up against unfamiliar disciplines and traditions is a way to create new research questions, approaches, and insights.

This is not to argue that we have found a silver bullet for interdisciplinary research and collaboration but rather that opening spaces for those encounters to occur is important. We have attempted to do so by supporting and organizing workshops, seminars, and writing retreats in which researchers from the different departments and faculties have been able to meet and interact. The most important initiative, however, has been the establishment of a university-wide smart city PhD network. At the launch of the network, seven PhD positions related to smart cities were announced – approximately one for each thematic topic and located across four faculties. These PhD positions were the core of the newly established PhD network where, from the beginning, we organized seminars and workshops focused on interdisciplinary research and collaboration and on creating a social environment. Most of the PhD students come from international backgrounds and only moved to Stavanger for the PhD position, with little to no social networks in the region. Furthermore, several students arrived in Stavanger only weeks before the first COVID-19 lockdown in the spring of 2020 and a few were stranded in their home countries and had to postpone the beginning of their studies. Thus, in the beginning many events took place online, which was far from ideal.

The network has thus based activities grounded in the firm belief that interdisciplinary engagement and collaboration is about being open-minded, but it also requires trust and familiarity, especially in the early phases of a process. The purpose of the PhD network has been to encourage and foster a cohort of researchers who are comfortable and open to interdisciplinary engagements and getting to know each other, not only as researchers but also as people. For this reason, we have organized dinners, study trips to other European cities, and writing retreats where students can interact professionally as well as socially across departments and faculties. The network has proved quite popular and now counts 27 affiliated PhD students from four different faculties.

Furthermore, to foster interdisciplinary understanding and collaboration, a PhD course was set up for the spring of 2020. The PhD course is a 10 ECTS (European Credit Transfer and Accumulation System) course on smart cities as an interdisciplinary research field and meant to encourage students to reflect on *smart city* as a concept that makes connections to various dimensions of sustainability. The learning outcome of the course is to enable students to position their own work and research in relation to dominant smart city discourses and theories. Finally, the course aims at enabling the student to discuss and develop interdisciplinary concepts, methodologies, and tools for smart city research. The course is organized such that the students are tasked with presenting their research to their peers who are not from the same discipline. Two students from other disciplines are asked to provide constructive suggestions on how perspectives from their field can enrich the research project. The PhD network activities all aim to bring together researchers in training working on a similar topic to discover synergies between different disciplines and approaches but also, and just as important, to show how no single discipline can provide the complete answer to the questions we are pursuing. Thus, knowing the limits of one's own discipline(s) and approach(es) is central to trans- and interdisciplinary collaboration.

We have also sought to engage PhD students and private and public sector organizations early in the PhD training. Most of our PhD students are from abroad and thus have little knowledge of what is going on regionally. Informing students about regional economic development strategies and initiatives and connecting them with relevant stakeholders have been important, both for the PhD students and to show the relevance of our research activities to the regional partners.

Working across sectors

Tackling urban challenges also requires cross-sectoral collaboration. This has been a central tenet of the research network as a whole from the outset. Neither academia, the public sector, the private sector, nor civil society can solve these challenges alone. There is a need to cultivate a sense of humility about one's own sector's ability to solve the complex challenges facing cities today and in the future. This in turn requires an awareness of the limitations of each sector and group, as well as respect for the contributions of others. Hence, knowledge and respect for the different areas of competences and responsibilities across different sectors are important. There are, for example, often pre-conceived notions about academics being too far removed from practice, while academics often lament that practitioners jump to solutions all too quickly while overlooking the theoretical and evidence-based foundation. These are partly true but, in my experience, also often exaggerated differences that can be overcome through dialogue.

This does not mean that cross-sectoral collaboration is easy. Here I wish to highlight two challenges that we have faced in enabling cross-sectoral collaboration. The first aspect to highlight is temporal. Practitioners often look for concrete solutions to problems in the now. Researchers, on the other hand, often work in a different temporal space looking either forward or backward in time. Researchers working on new technologies often focus on emerging technologies; that is, solutions that are not commercially available, and perhaps years from commercialization. These are often of little relevance to here-and-now problems. Similarly, social science and humanities researchers studying, for example, the implementation of new technologies do not have their research findings ready for dissemination until long after the implementation has ended – and by then practitioners have already moved on. This temporal asynchronicity must be addressed at both ends.

The second aspect is a matter of co-designing projects or interventions. Quite often, an initiative is started by smaller groups that usually consist of perhaps partners from one or two sectors who define the project scope. Partners from other sectors are invited in later in the process. This may create tensions if it becomes a repeat pattern. My colleagues in social science disciplines often lament the fact that they are added on at the end of a technical project development process to 'address the societal aspects'. Collaboration between partners must necessarily be more equal and inclusive from the beginning, not only for fairness but, more important, because understanding the multiple dimensions of a problem is necessary before moving to devising potential solutions.

Building regional collaborative relations across sectors has been an essential part of our activities. In this regard, we have been lucky to have a relatively easily legible landscape of players. The network of regional smart city municipalities led by the City of Stavanger Smart City Office has held almost weekly meetings to discuss municipal challenges and development opportunities. The private sector has been mainly organized through the Nordic Edge Smart City Innovation Cluster. With the launch of the Research Network for Smart Sustainable Cities at the University of Stavanger, collaboration between academia, the public sector, and the private sector has become easier. These three networks provided easier access points for collaboration. No longer did municipalities or companies have to scroll through the academic profiles of hundreds of researchers at the university and, similarly, researchers no longer had to identify which department in the municipality would be the most relevant to connect with.

Regular meetings and activities between the different networks have been instrumental in building connections and getting to know each other. We are not always aligned, but when mutual interests have been identified, these points of contacts have enabled relatively quick mobilization. Being in a medium-sized urban region has also helped a lot. The number of actors is smaller than in a big metropolitan area, so after a few years, most key actors know each other quite well. We have held multiple workshops and seminars throughout the years in which researchers, entrepreneurs, and public sector employees have been able to meet, connect, and discuss possible collaboration. During the COVID-19 lock-downs, we organized several online Science Talks together with Nordic Edge in which researchers presented their research followed by an online panel discussion with representatives from the public sector and industry. These events have led to multiple new collaborations.

One area of collaboration that has been particularly successful has been around student involvement. We quickly realized that students' interests in working on concrete challenges aligned well with the needs of the public sector and companies. Nordic Edge Innovation Cluster thus took the lead in developing a range of activities targeting students. These include matching students looking for bachelor's and master's thesis topics with the needs of industry and public sector and offering student internships and student jobs in both the private and public sectors. These activities have been immensely popular among our students. At the latest 'Find Your Thesis' event, 22 companies came to the university and more than 100 students showed up to see whether there were challenges that matched their interests. Similarly, the 'Join a Startup Night' events organized by Nordic Edge usually attract more than a hundred students interested in exploring employment opportunities with startups in the region.

In the past 4 years, collaboration with the public and private sectors within the smart city space has expanded significantly. It is not always easy, and sometimes frustrations occur, but, overall, all three sectors see the benefit of closer collaboration. When it comes to civil society engagement, the challenges have been greater. There are not many civil society organizations and citizens who are particularly engaged with the *smart city* as a concept. As one journalist in the local newspaper argued in an editorial in 2019: 'A smart city must be relevant to its residents' (Øvrebekk, 2019). A survey made by the Norwegian Ministry of Local Government and Modernisation in 2019 showed that citizen involvement in smart city initiatives and projects in Norway is quite limited (Stenstadvold, 2019). This may be due to governance structures, but as the journalist alluded to in the editorial, perhaps it is also the nature of many smart city projects that mainly focus on technological innovations to optimize traditional municipal services such as waste management, street lighting, and flood control. The smart city, it appears, is simply too far removed or perhaps too abstract to arouse the interest of citizens.

Engagement with civil society will probably require a shift in which topics are addressed to make smart city initiatives more relevant to everyday lives. Furthermore, there is probably also a need to reconsider when and how citizens are involved. The previously mentioned report from the Norwegian Ministry of Local Government and Modernisation clearly highlights that citizen involvement mainly occurs through hearings, workshops, needs assessments, feedback sessions, and so forth but that citizens or civil society rarely are involved at the strategic levels. This is recognizable from Stavanger as well. The smart city agenda is mainly set by municipal and private sector interests. To be fair, the smart city office in Stavanger has sought to reach out to citizens in various ways, but citizen interest and engagement remain low. This is an area that will require increased focus in the coming years, and perhaps the discursive shifts towards climate neutrality and inclusion may prompt more citizens and civil society organizations to get involved in urban development. This remains to be seen.

What did the smart city do to us?

I can only speak for myself, so the following reflections on what the smart city agenda has done to the region of Stavanger is not representative, yet I do think there are lessons to be learned more broadly. My experience is that in the past 4 years, the collaboration between the university research network and the private and public sectors has improved significantly. Yet there are still many ways in which interdisciplinary and intersectoral collaboration can be improved. Institutional and intersectoral silo thinking remain important obstacles to addressing the complex urban and regional challenges that the smart city agenda is supposed to address. Forging connections and collaboration across disciplines and sectors is an uphill struggle if the organizational and institutional structures still emphasize thinking and working in silos. To bring down those barriers, a first step is to forge connections across disciplines and sectors.

When it comes to regional economic development from smart city initiatives, it is difficult to assess the impact. At the time of this writing, Stavanger is experiencing a boom because of rising oil and gas prices, attracting workers who may otherwise seek opportunities in the smart city space. Nevertheless, we have demonstrated some success in attracting new talent to start-up companies thanks to the collaboration with Nordic Edge Innovation Cluster. But whether the smart city constitutes a potential pillar of the future regional economy remains to be seen. The regional development strategy based on smart city innovation remains more of a promise than a fact of economic development and innovation. When it comes to public sector innovation and change, the smart city agenda is not something that permeates the organizational structures and cultures. Those working with smart city initiatives remain single individuals in the smaller municipalities or, as in Stavanger, working in separate departments. Yet, we have also managed to build collaboration and projects between sectors. Perhaps the most tangible outcome is an increasing ability to quickly mobilize and organize project consortia when needed for concrete objectives. It no longer takes months to identify the right people and departments when opportunities for project development arise. It is also my experience that we have become better at involving other colleagues and other sectors earlier in the project development phase and to remember to include partners from different departments and sectors in project development.

Reflecting on what the smart city agenda has done to us, one main effect is perhaps increased dialogue and discussion across organizational and institutional boundaries. This has also led to many discussions on the substance of the smart city agenda. Over the past 4 years, citizen engagement has become more important, pushing back against the emphasis on technological innovation in smart city projects. Furthermore, as the European agenda has shifted from smart cities towards climate-neutral cities, most notably through the EU missions, sustainability and climate neutrality have become more prominent agendas. This is also reflected in the university's research network, which in 2021 changed its name to add *sustainability* to better reflect the scope of our activities. One could argue that the terms *smart* and *sustainable* are vague and lacking clear definitions. To some this may appear as a weakness, but from the perspective of attempting to build collaboration across disciplines and sectors, such vague terms may be what enable dialogue and cooperation, allowing a broad spectrum of stakeholders to engage (although citizen involvement remains lacking) in initiatives. We may not all agree on what exactly smart and sustainable entail in terms of substance, but they do allow us to engage in different and varied ways without having to agree on everything before moving ahead. This may prove a challenge to some academics, but bridging theory and practice sometimes requires compromises to foster change.

Note

1　Sabadell (Spain), Leipzig (Germany), and Prague (Czech Republic) were included as follower cities intended to learn from the experiences of the three main cities. Tianjin (China) was also part of the project as observer city.

References

Asle Bergsgard, N., & Vassenden, A. (2011). The legacy of Stavanger as Capital of Culture in Europe 2008: Watershed or puff of wind? *International Journal of Cultural Policy*, 17(3), 301–320. doi:10.1080/10286632.2010.493214

Kongsnes, E. (2015, August 8). *NAV: 36.357 jobber er kuttet.* aftenbladet.no. https://www.aftenbladet.no/aenergi/i/3530X/nav-36357-jobber-er-kuttet

Marco-Serrano, F., Rausell-Koster, P., & Abeledo-Sanchis, R. (2014). Economic development and the creative industries: A tale of causality. *Creative Industries Journal, 7*(2), 81–91. doi:10.1080/17510694.2014.958383

Munkvik, C. (2014, October 7). *Stavanger blir europeisk fyrtårn for smarte byer.* aftenbladet.no. https://www.aftenbladet.no/lokalt/i/mL9Av/stavanger-blir-europeisk-fyrtaarn-for-smarte-byer

Nerheim, G. (2014). Oil shocks in an oil city: The view from Stavanger, Norway, 1973–2008. In J.A. Pratt, M.V. Melosi, & K.A. Brosnan (Eds.), *Energy capitals: Local impact, global influence* (pp. 127–144). University of Pittsburgh Press. doi:10.2307/j.ctt6wr9s3

Nygaard, E. (2009, April 16). *Energetic Stavanger – An energy capital rooted in success with a view to the future.* Norway Exports. https://www.norwayexports.no/news/energetic-stavanger-an-energy-capital-rooted-in-success-with-a-view-to-the-future/

Øvrebekk, H. (2017, September 28). *«Stavanger kan bli stor på smart-by-teknologi i Kina».* aftenbladet.no. https://www.aftenbladet.no/aenergi/i/8evKd/stavanger-kan-bli-stor-paa-smart-by-teknologi-i-kina

Øvrebekk, H. (2019, September 24). *Ut med teknologi, inn med mennesker.* aftenbladet.no. https://www.aftenbladet.no/meninger/kommentar/i/QoOXLR/ut-med-teknologi-inn-med-mennesker

Stavanger Aftenblad. (2016, September 29). *2 Millioner i støtte til Nordic Edge Expo.* https://www.aftenbladet.no/lokalt/i/p5xKW/2-millioner-i-stoette-til-nordic-edge-expo

Stavanger Municipality. (2016). *Roadmap for the Smart City Stavanger.* https://www.stavanger.kommune.no/siteassets/samfunnsutvikling/planer/engelske-planer/roadmap-smart-city-stavanger-2016.pdf

Stenstadvold, M. (2019). *Smarte byer og kommuner i Norge – En kartlegging.* Kommunal- og Moderniseringsdepartementet.

Triangulum. (2014). *Triangulum.* https://triangulum-project.eu/

Part II
The conceptual smart city

3 Tensions and opportunities for cross-disciplinary collaboration in smart city work

Stella Huang, Reidar Staupe and Barbara Maria Sageidet

Introduction

The importance of cross-disciplinary work is increasingly highlighted. As Barthel and Seidl (2017, p. 305) pointed out, 'To effectively tackle the major challenges facing society – such as energy, water, climate, food, and health [researchers] must work collaboratively across multiple fields'. While disciplinary research continues to generate valuable insights, the diverse perspectives and methodologies employed by different disciplines pose significant challenges (Nissani, 1997; van Rijnsoever & Hessels, 2011).

As the challenges of modern life continue to grow in complexity, cross-disciplinary collaboration has become increasingly prevalent in the field of the smart city (Bibri, 2018a, 2018b, 2019). Correspondingly, realities in a smart city – for example, in Stavanger – demonstrate the challenges of cross-disciplinary research. Bibri (2018a) underlined the importance of bringing together multiple disciplines in the service of smart cities but also argued for the need to move beyond the idea of academic disciplines rather than speaking of collaboration between these. However, despite increasing attention to the importance of cross-disciplinarity, there is a need for a better understanding of how such integration might take place and which challenges are involved (Mora et al., 2022). We argue that cross-disciplinary research is essential for advancing knowledge and developing practical solutions for smart cities. However, it requires a deep understanding of the complexities of cross-disciplinary collaboration and the awareness of the tensions between scientific, environmental, social, institutional, and political practices, to promote the development of sustainable systems in the smart city, in line with the global Sustainable Development Goals (SDGs; United Nations, 2015).

This study sets a focus on the perspective of early- and mid-career researchers, to exemplify how disciplinary dynamics of collaboration may be experienced. While both groups encounter similar challenges, mid-career professionals tend to reflect on more distant experiences. Although a heterogenous group, early- and mid-career professionals and researchers are usually highly skilled or qualified, digitally literate individuals (Curşeu et al., 2021). They are often young, mobile, creative, and interested in networking and collaboration and contribute essentially to an innovative and expanding knowledge-based economy (Bibri, 2018b).

DOI: 10.4324/9781003498650-6

A focus mapping literature review (Bradbury-Jones et al., 2019) is used to explore the challenges and tensions that arise from the differences between disciplines and from integrating them as part of cross-disciplinary research. Further, three interviews were carried out to enrich the analysis with perspectives from personal experiences. Specifically, we aim to answer the explorative questions: What are the challenges entailed with cross-disciplinary research? How may these challenges be overcome with special regard to early-career researchers and to the smart city?

Finally, the study reflects on possibilities to overcome identified challenges and on potentials we may find in the smart sustainable city, to facilitate multi-, trans-, and interdisciplinary collaborative approaches, especially for early-career scholars. This chapter will thus serve as food for thought for practitioners, early-career, mid-career, and seasoned scholars alike who co-produce cross-disciplinary works within the smart city context.

The smart city as a field of collaboration

Research on smart cities has been extensive (Anthopoulos, 2015) and has evolved to encompass environmental sustainability, as cities confront rising environmental and social challenges while focusing on improving the quality of life, infrastructure, mobility, energy and other resources, and economic viability of urban areas through innovative technological solutions (Bibri & Krogstie, 2017; Curșeu et al., 2021; Duygan et al., 2022; Reid et al., 2010). This growing research field requires the collaboration of various disciplines to solve real-world problems, including urban planning, engineering, computer science, economics, social sciences, public health, risk evaluation, climate adaptation, and environmental sciences (Bakıcı et al., 2013; Batty, 2013; Chatterjee et al., 2018; Cruz-Jesus et al., 2017; Liu et al., 2012; MacDonald et al., 2022; Washburn & Sindhu, 2010; Zygiaris, 2013; cf. United Nations, 2015).

Academic disciplines and forms of collaboration

The concept of *discipline* itself lacks clarity as established disciplines vary widely in scope and epistemological tradition (Staupe-Delgado et al., 2022). A discipline denotes a tradition of scientific knowledge production with each discipline developing theories and methods tailored to its chosen area of study (Newell, 2001). Such disciplines function as cultural systems organizing epistemological communities and play a significant role in identity formation (Pavalko, 1988). They facilitate the consolidation of cumulative knowledge, often resulting in 'paradigms', as described by Kuhn (1970). However, while cross-disciplinary collaboration has the potential to enhance innovative solutions, it can also serve as a barrier due to differences in assumptions about the world, criteria for knowledge, identities, vested interests, and traditions (Lattuca, 2002). In essence, it is crucial to recognize that reality is multifaceted, interrelated, and complex, often defying reduction to a singular approach to knowledge production and problem solving (Klein, 2005). According to Staupe-Delgado and colleagues (2022, p. 2):

> It is difficult to discern the point at which an academic field becomes a discipline. [...] Those who defend the rather hierarchical understanding of scientific knowledge production claim that at the highest (or broadest) level, there are 'sciences' like natural and social sciences. These, in turn, can be divided into parent, root or reference 'disciplines' like mathematics or psychology, which in turn can be divided into 'subdisciplines', such as civil engineering or political psychology. Some of these can then be called 'fields' which, while relying on their parent disciplines, have to strive for their own identity.

From a relational perspective, multi-disciplinary fields evolve as people originating from different disciplinary boundaries join forces to address multifaceted societal, environmental, and technical challenges, also highlighting the need for smart cities (Klein & Newell, 1997). From a power-sensitive perspective, participants in such issue domains make claims about how they have come to constitute a field or an emerging discipline. Whether and when a field has reached the status of 'emerging discipline' is debatable, while an increasing number of research areas – for example, 'sustainability science' (Komiyama & Takeuchi, 2006), 'leadership studies' (Riggio, 2011), 'safety science' (Wang et al., 2020), and 'energy transitions studies' (Araújo, 2014) – seem to pursue this status of discipline. However, even if such emerging disciplines seem to meet the above definitional criteria, such as having developed distinct vocabularies, approaches, and methods (Staupe-Delgado et al., 2022), they can later become objects of controversy as more established disciplines may challenge their claims. Across a range of societal challenges, diverse disciplinary perspectives are put to work and synthesized into comprehensive frameworks, fostering holistic understandings of urban complexities but also frictions, as well as external critique (Bibri, 2018b). Such critique is arguably generative of overall scientific progress; for example, in terms of paradigm shifts in the Kuhnian sense (Kuhn, 1970).

There are also strengths to such plurality and contradictions, however. As Klein (2005, p. 45) suggested, 'Difference, tension, and conflict are not barriers that must be eliminated. They are part of the character of interdisciplinary knowledge negotiation'. Cross-disciplinary collaboration thus serves to soften up paradigms and support the building down of conceptual siloes. Other forms of knowledge production rely more explicitly on problematization and critique, which are also recognized as serving a vital function for correcting potentially harmful or problematic applications of knowledge (Alvesson & Sandberg, 2011). Nissani (1997) highlighted that errors from disciplinarians may be detected by people with cross-disciplinary knowledge, which also often promotes creativity.

Following Petts et al. (2008), Menken and Keestra (2020), and Staupe-Delgado et al. (2022), we can distinguish between several different approaches of working across disciplines:

- *Multi-disciplinarity* refers to multiple disciplines coming together to work on an issue, while each discipline contributes its perspective individually without integration. Example: A software developer and a geographer make a smartphone application together. A geographer contributes with knowledge of spatial behaviour, and the software developer designs an application based on this knowledge, without either person engaging meaningfully in the expertise of the other.

- *Inter-disciplinarity* denotes the integration and mixing of different approaches, methods, and insights from a range of disciplines to address a problem. Example: People with various disciplinary backgrounds come together to improve an issue such as urban mobility, and this collaboration is of such a nature that the disciplines come together in a coherent framework, in a product or outcome that no longer belongs to any specific discipline.
- *Intra-disciplinarity* is when a single discipline or field of study reflexively examines its own methods, state of knowledge, strengths, limitations, and progress. One example is how a field will often eventually start to question its own status as an emerging discipline, just like this very chapter is doing for smart city research.
- *Trans-disciplinarity* is often taken to connote an approach that goes beyond a sole focus on academic disciplines to draw on other forms of expertise, such as lay knowledge, decision makers, and other experts, to solve societal problems (see also Lang et al., 2012). Example: A range of scientists from different disciplines teams up with artists, civil society groups, the municipality, and local volunteers to address a problem like gentrification (see also Lorentzen and Langhelle, chapter 4).
- *Post-disciplinary* approaches challenge the notion of academic disciplines altogether and represent a shift beyond traditional disciplinary structures. Post-disciplinarity represents a more radical approach centred on understanding and/or addressing societal challenges with less concern for academic conventions and traditions. Example: Smart city researchers may start to identify as smart city researchers, dedicated to overcoming smart city challenges, and may lose whichever disciplinary identity they might initially have had when coming into the field. Likewise, we may gradually see more full-fledged doctorates or professorships in 'smart city'.

Methodology

This chapter is based on a mixed methods approach, combining a literature review and interviews. The study builds on a focus mapping review (Bradbury-Jones et al., 2019), an approach to investigating trends and what is happening within a particular subject or field during a time to be screened. This study's review was conducted between October 2021 and January 2022 and reviewed a time frame of 20 years (2002–2022) corresponding to a period when cross-disciplinary city services were globally developing towards more efficient and flexible response to urban needs (Kim, 2022).

In addition to the literature review, three semi-structured interviews were conducted, inspired by Horton et al. (2004), see Figure 3.1. Two interviews were conducted to provide exemplified insights into the real-world experiences of an anonymous early-career scholar ('Jennifer') and an anonymous mid-career scholar ('Gavin'). Further, Professor Klaus Mohn, head of the University of Stavanger, was interviewed as representing a stakeholder to provide links between cross-disciplinarity, early-career scholars, and the smart city.

Literature review

The first phase of the literature review followed a three-step process (Bradbury-Jones et al., 2019), using the search engines Google Scholar and Web of Science; see Figure 3.1. The search strategy employed a broad Boolean search with terms like 'interdisciplinary', 'challenge', 'smart city', and related keywords. This approach allowed for the identification of relevant literature across a wide range of disciplines, ensuring that the review captured the diverse nature of cross-disciplinary challenges.

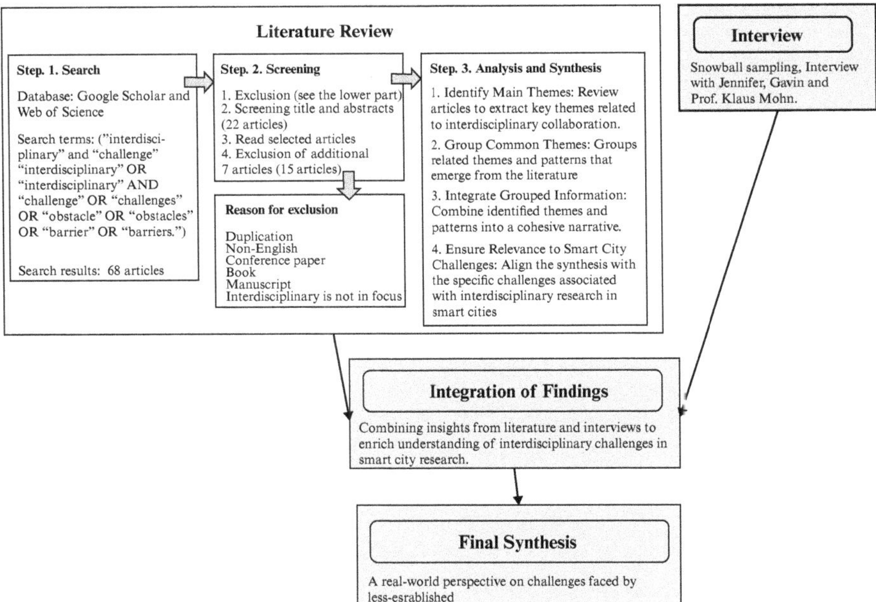

Figure 3.1 Research methods

Analysis and integration of findings

The initial search yielded 68 articles. These were further screened for relevance, based on their titles and abstracts. A special focus was set on studies addressing interdisciplinary challenges and potential solutions, while studies addressing specific disciplines or those that did not directly engage with interdisciplinary challenges were excluded. To further refine the selection, to increase relevance and reduce redundancy, articles containing repetitive information were filtered out, with a preference given to original sources or review articles that offered unique or foundational perspectives on cross-disciplinary challenges. This refined screening retained 22 articles, and their full-text review led to the exclusion of 7 additional articles.

In the final selection of 15 articles, common themes and patterns were identified and grouped in relation to the research questions. The focus and scope of each article are summarized in Table 3.1.

Analyses were further accomplished through the construction of a coherent qualitative synthesis (cf. Figure 3.1) or narrative (Chase, 2013), summarizing the key findings, methodologies, and recommendations from the 15 selected articles, together with key contents from the interviews, to address the research questions, while being aware of influences by the author's preunderstandings, beliefs, and biases.

Findings and discussion

What are the challenges entailed with cross-disciplinary research?

Despite of the focus mapping, challenges entailed with cross-disciplinarity research are only weakly in focus in many of the selected articles. However, they reveal that researchers face numerous challenges when confronting cross-disciplinary research, from publication hurdles to career advancement issues. A frequently addressed challenge to interdisciplinary work is that it often involves integrating what are fundamentally distinct philosophical approaches and intellectual traditions. As Petrie (1976, p. 11) wrote, 'Quite literally, two opposing disciplinarians can look at the same thing and not see the same thing'. Problems associated with collaboration are not about data or research design but perceptions of the world at a higher level of abstraction, related to different ontological and epistemological assumptions and paradigms that are difficult for individuals to simply put aside.

Traditional academic degrees and assessment criteria predominantly favour mono-disciplinary approaches, compelling many interdisciplinary researchers to align their profiles within a single discipline (van Rijnsoever & Hessels, 2011). Fixed-term contracts and short-term evaluations do not value the long-term efforts required for interdisciplinary contributions. Conflicting epistemological perspectives further complicate the recognition of interdisciplinary work, with critics arguing that interdisciplinary researchers may lack depth in specific academic contexts despite providing innovative insights in thematic areas. Scepticism towards interdisciplinary collaboration remains an obstacle, hindering success (Schuitema & Sintov 2017).

Academic institutions are primarily organized around disciplinary lines, posing additional challenges for interdisciplinary researchers (Schuitema & Sintov, 2017). Early-career researcher Jennifer highlights the difficulty of moving beyond disciplinary silos, stating, 'It's all well and good to have the best intentions to work interdisciplinary, but oftentimes, folks end up stuck in their own little siloes. They

Table 3.1 An overview of the articles that mention challenges in cross-disciplinary studies

Author (year)	Focus	Scope	Whether mentioned early- or mid-career researcher
Bammer (2016)	The focus of this article is on improving the effectiveness of peer review for inter-disciplinary research proposals	Peer review	
Bibri and Krogstie (2017)	A comprehensive overview and critique of smart sustainable cities	Smart and sustainable city	
Brister (2016)	Explore how epistemic differences between disciplines can hinder inter-disciplinary collaboration	Debates between conservation biologists and anthropologists	
Chmutina et al. (2022)	The authors will discuss their recent experiences with the peer review process in a highly interdisciplinary field, namely, disaster studies, and demonstrate how debate can be easily suppressed	Peer review	
Forsberg et al. (2022)	Presenting various peer review practices and examining various evaluative activities in contemporary academic settings	Peer review practices Education	
Haklay (2017)	Exploring how the smart city paradigm shapes the urban environment and societal interactions	Smart and sustainable city	
Israilidis et al. (2021)	This is a review of the existing smart city literature and integrates knowledge perspectives to provide an overview of future research directions	Smart city research	
Pautasso and Pautasso (2010)	This article provides an overview of questions related to the peer review process of interdisciplinary manuscripts	Peer review	

Author (year)	Focus	Scope	Whether mentioned early- or mid-career researcher
Schuitema and Sintov (2017)	This study highlights the need to address institutional barriers and research challenges and provides recommendations for overcoming them	Energy research	x
Shell et al. (2021)	Presents challenges related to self-driving autonomous vehicle (AV) shuttles in the smart city context	AV project	
Szostak (2002)	The focus of this study is to develop a guide for conducting interdisciplinary research	Interdisciplinary research	
Thompson et al. (2017)	The perspectives of stakeholders and scientists involved in a transdisciplinary research project	Transdisciplinary research in smart city	x
van Noorden (2015)	This paper uses numbers to reveal the growth and influence of interdisciplinary research	Interdisciplinary research	
van Rijnsoever and Hessels (2011)	Investigating the characteristics of researchers associated with disciplinary and interdisciplinary research collaborations	Interdisciplinary and disciplinary research	
Weichselgartner and Truffer (2015)	This report introduces the concept of smart city, collection of infrastructure and services in smart cities	Smart city	

stick to the methods and theories within their own field – the ones they're most comfortable and familiar with'. Additional administrative and teaching duties impose significant time constraints. The structure complicates defining positionality for those not aligned within a single discipline.

Brister (2016) highlighted the importance of clear communication between collaborators to avoid disciplinary capture, which was underlined by Nissani (1997). However, amidst the multitude of academic responsibilities, prioritizing communication can be daunting. Communication within interdisciplinary teams is absolutely crucial; according to Jennifer,

> But here's the thing – it's not always easy to find the time for it. … It can be tough to prioritize communication. And without constant support and resources, it's all too easy for people to drift back to their own field. Breaking out of that bubble can be a real uphill battle.

Without continued support and resources, individuals risk reverting to their comfort zones, impeding the progress of interdisciplinary collaboration (McNair et al., 2015).

Pautasso and Pautasso (2010) noted the difficulties interdisciplinary researchers face in publishing in highly ranked journals, which often select reviewers who do not possess sufficient expertise in interdisciplinary integration (Schuitema & Sintov, 2017). This issue is exacerbated by the 'blind' review process (Forsberg al., 2022), while specialization aids editorial staff in identifying suitable reviewers, potentially excluding interdisciplinary perspectives (Chmutina et al., 2022). Projects that are too cross-disciplinary can be penalized based on evaluator criteria. Journals typically favour projects within single disciplines, making it tough for interdisciplinary works to gain recognition (Rylance, 2015), particularly in natural and life sciences. These challenges extend to broader academic evaluations, including funding proposals and tenure decisions. While funding bodies sometimes support collaboration across disciplines, they often use disciplinary categories for evaluation panels, creating a mixed incentive structure (Rylance, 2015). Despite this, the citation impact of interdisciplinary publications has been increasing, indicating long-term benefits (van Noorden, 2015). Also, interdisciplinary publishing in disciplinary-oriented outlets is becoming more common and promotes assimilation of research findings across fields.

While epistemic standpoints and methodological conflicts between disciplines can initially seem insurmountable (Brister, 2016; Franks et al., 2007; Lang et al., 2012), these challenges also present opportunities. Interdisciplinary research is well-suited to empirically investigate the coherence and logical consistency of different disciplines (Szostak, 2002). By introducing external theories and methods, interdisciplinary approaches can reveal overlooked aspects, challenge established assumptions, and suggest alternative explanations that may offer better explanatory power (Nissani, 1997). This process not only enhances our understanding of individual disciplines but also identifies areas where disciplinary perspectives may be incomplete or even incoherent.

Nevertheless, the temptation to compromise the quality of interdisciplinary knowledge often looms, as it can be easier to default to safe, uncontroversial topics – commonly known as 'low-hanging fruit research' (Freeth & Caniglia, 2020; McNair et al., 2015; Weichselgartner & Truffer, 2015). The true art of interdisciplinary research lies in researchers confronting the inherent tensions of collaborative work, while both articulating and integrating diverse perspectives (Thompson et al., 2017). Therefore, it is imperative for institutions to prioritize the provision of resources and opportunities for interdisciplinary training, collaboration, and research.

How may these challenges be overcome with special regard to early-career researchers and to the smart city?

Jennifer notes the struggles to stay updated in one's own field while engaging in cross-disciplinary work, stating, 'With the increasing amount of administrative tasks, teaching, and student-related work, it can be tough to have time left for research and staying up-to-date', she often feels overwhelmed by the volume of new information. Particularly, early- and mid-career researchers face limited incentives and extra costs for interdisciplinary research, as fixed-term contracts and short-term performance metrics hinder interdisciplinary career paths (Rhoten & Parker, 2004). Despite these obstacles, trends are moving towards greater facilitation of interdisciplinary work.

Only two of the selected articles (see Table 3.1) specifically address challenges of interdisciplinary research in the context of smart cities. While Haklay (2017) stressed the importance of human and environmental values in the smart city, Bibri and Krogstie (2017) underlined that the smart city may encourage interdisciplinary collaboration and innovation through the exchange of ideas and approaches from different disciplines in an applied practical context. While the smart city creates an ecosystem that fosters cross-disciplinary collaboration and innovation (Curşeu et al., 2021), a challenge may be the need for deep understanding of these complexities. Also, awareness of the tensions between scientific, environmental, social, institutional, and political practices is crucial to develop sustainable systems in the smart city. Bibri (2018a, 2018b) highlighted that the smart city may attract early- and mid-career professionals. Through concerted efforts, institutional hubs can empower these professionals to navigate interdisciplinary research, and they can nurture interdisciplinary talent and foster collaborations. Multi-disciplinary, transdisciplinary, and interdisciplinary teams can identify and address interconnected challenges more effectively by drawing on diverse perspectives and expertise (Bibri, 2018a).

There is a noticeable surge in commitment across various collaborative institutions, training programs, projects, and research initiatives, particularly focusing on sustainable transitions (United Nations, 2015). Also, funding calls in the EU increasingly demand interdisciplinarity, particularly for smart city and green transition themes. The Stavanger Triangulum project (including its challenging) is an example (see Aunemo et al., chapter 1, Müller, chapter 2, and Nencioni et al., chapter 11).

However, before potentially investing in an interdisciplinary project, funding agencies generally demand a solid basis in existing research and a compelling narrative and record showing a high probability of success (Bammer, 2016). The rector of the University of Stavanger in Norway, Klaus Mohn, suggests that good proposals should build on diverse research resources and provide a diversity of spaces for participation. Proposals should utilize the competences of both experienced and early-career researchers. He further underlines that young or early-career researchers usually contribute with innovative ideas, courage, and intellectual force (cf. Nissani, 1997). According to Shell et al. (2021), challenges can be overcome by cultivating an environment that values interdisciplinary contributions and provides resources for training and collaboration. As the mid-career researcher Gavin aptly puts it, 'It's important that we produce both in-depth and interdisciplinary research, depending on how to ensure results and insights that we wouldn't otherwise obtain'. Klaus Mohn also emphasizes the necessity of balancing in-depth research through multi-disciplinarity with interdisciplinary work to foster innovation. Thompson et al. (2017) proposed encouraging networks of participants (professionals, stakeholders) to shape their own appropriate working network, including space for multi-, trans-, and interdisciplinarity.

Early-career researchers may have in mind how certain educational settings are beginning to embrace thematic positions advertised within smart cities. This is evident in emerging urban-centric subjects within academic institutions such as the University of Stavanger (UiS, 2024) and the Norwegian University of Science and Technology (n.d.), both in Norway, and Linköping University (n.d.). This trend is fuelled by mandates from governmental bodies and research councils, emphasizing academia's responsibility to drive significant societal change and transitions to contribute to the achievement of the SDGs (United Nations, 2015).

Klaus Mohn underlines that the mindset of the smart city has influenced the programmes and courses at the UiS and has made them more relevant to the society, stating, 'Relevance is a crucial part of quality'. Further, Mohn highlights that UiS has funded PhD projects for young researchers within smart city themes as part of the university's aim towards sustainable development of the region, its infrastructure, technology, and energy sector.

Finally, communication and knowledge exchange are crucial to inherently overcome challenges related to effective interdisciplinary collaboration and leads to novel technologies, policies, and strategies to enhance urban sustainability, resilience, and liveability (Bibri, 2019; Israilidis et al., 2021). Mohn underlines that UiS and the smart city needs researchers who are willing to build the competences needed to solve problems to promote the green shift and the SDGs (United Nations, 2015), including single disciplines, as well as multi-, trans-, post-, and interdisciplinary approaches.

Conclusion and recommendations

Challenges in urban societies span across various disciplines. Addressing these complex problems requires dynamic efforts across disciplines. This chapter identified challenges but also opportunities for early- and mid-career professionals/researchers that may be especially prevalent in smart cities. Namely, in smart cities these researchers and professionals are needed as a pool for specialized expertise and research skills, typically paired with youth, openness, creativity, mobility, flexibility, and eagerness for international networking and collaboration, all attributes that are vital for fostering innovative interdisciplinary collaborations.

Stakeholder, employers, universities, publishers, and other institutions should contribute to break down barriers and promote various forms of cross-disciplinary research, as they appear necessary in critical areas of a smart city. We recommend funding agencies to establish mechanisms and criteria to specifically support research activities across disciplines and to encourage and specifically give value to the inclusion of less experienced researchers and professionals.

Universities and academic communities must take proactive steps to foster interdisciplinary research. Grants or scholarships for interdisciplinary courses, workshops, training, and mentorship programs, especially for students and young researchers, can help to develop the necessary skills for effective collaboration. Interdisciplinary experts, or those from multiple disciplines, should review and evaluate submissions and proposals, to strengthen an inclusive peer review processes and to ensure accurate assessment of interdisciplinary integration. Collaboration across different fields is essential, as it leverages diverse expertise and perspectives, strengthening the research's quality.

References

Alvesson, M., & Sandberg, J. (2011). Generating research questions through problematization. *The Academy of Management Review*, 36(2), 247–271. doi:10.5465/AMR.2011.59330882

Anthopoulos, L. (2015). Understanding the smart city domain: A literature review. In M. Rodríguez-Bolívar (Ed.), *Transforming city governments for successful smart cities*. Springer. doi:10.1007/978-3-319-03167-5_2

Araújo, K. (2014). The emerging field of energy transitions: Progress, challenges, and opportunities. *Energy Research & Social Science*, 1, 112–121.

Bakıcı, T., Almirall, E., & Wareham, J. (2013). A smart city initiative: The case of Barcelona. *Journal of the Knowledge Economy*, 4(2), 135–148. doi:10.1007/s13132-012-0084-9

Bammer, G. (2016). What constitutes appropriate peer review for interdisciplinary research? *Palgrave Communications*, 2, 16017. doi:10.1057/palcomms.2016.17

Barthel, R., & Seidl, R. (2017). Interdisciplinary collaboration between natural and social sciences – Status and trends exemplified in groundwater research. *PLoS One*, 12(1), e0170754. doi:10.1371/journal.pone.0170754

Batty, M. (2013). *The new science of cities*. MIT Press.

Bibri, S.E. (2018a). Backcasting in futures studies: A synthesized scholarly and planning approach to strategic smart sustainable city development. *European Journal of Futures Research*, 6(13), 2–27. doi:10.1186/s40309-018-0142-z

Bibri, S.E. (2018b). A foundational framework for smart sustainable city development: Theoretical, disciplinary, and discursive dimensions and their synergies. *Sustainable Cities and Society*, 38, 758–794. doi:10.1016/j.scs.2017.12.032

Bibri, S.E. (2019). On the sustainability of smart and smarter cities in the era of big data: An interdisciplinary and transdisciplinary literature review. *Journal of Big Data*, 6, 25. doi:10.1186/s40537-019-0182-7

Bibri, S.E., & Krogstie, J. (2017). Smart sustainable cities of the future: An extensive interdisciplinary literature review. *Sustainable Cities and Society*, 31, 183–212. doi:10.1016/j.scs.2017.02.016

Bradbury-Jones, C., Breckenridge, J.P., Clark, M.T., Herber, O.R., Jones, C., & Taylor, J. (2019). Advancing the science of literature reviewing in social research: The focused mapping review and synthesis. *International Journal of Social Research Methodology*, 22(5), 451–462. doi:10.1080/13645579.2019.1576328

Brister, E. (2016). Disciplinary capture and epistemological obstacles to interdisciplinary research: Lessons from Central African conservation disputes. *Studies in History and Philosophy of Biological and Biomedical Sciences*, 56, 82–91. doi:10.1016/j. shpsc.2015.11.001

Chase, S.E. (2013). Narrative inquiry: Still a field in the making. In N.K. Denzin & Y.S. Lincoln (Eds.), *Collecting and interpreting qualitative materials* (4th ed., pp. 55–84). SAGE.

Chatterjee, S., Kar, A.K., & Gupta, M.P. (2018). Success of IoT in smart cities of India: An empirical analysis. *Government Information Quarterly*, 35, 349–361. doi:10.1016/j. giq.2018.05.002

Chmutina, K., Cheek, W., & von Meding, J. (2022). 'Critique is not a verb': Is peer review stifling the dialogue in disaster scholarship? *Disaster Prevention and Management*. Advance online publication. doi:10.1108/DPM-09-2021-0266

Cruz-Jesus, F., Oliveira, T., Bacao, F., & Irani, Z. (2017). Assessing the pattern between economic and digital development of countries. *Information Systems Frontiers*, 19(4), 835–854. doi:10.1007/s10796-016-9634-1

Curşeu, P.L., Semeijn, H.J., & Nikolova, I. (2021). Career challenges in smart cities: A sociotechnological systems view on sustainable careers. *Human Relations*, 74(5), 656–677. doi:10.1177/0018726720949925

Duygan, M., Fischer, M., Pärli, R., & Ingold, K. (2022). Where do smart cities grow? The spatial and socio-economic configurations of smart city development. *Sustainable Cities and Society*, 77, 103578. doi:10.1016/j.scs.2021.103578

Forsberg, E., Geschwind, L., Levander, S., & Wermke, W. (2022). Peer review in academia. In E. Forsberg, L. Geschwind, S. Levander, & W. Wermke (Eds.), *Peer review in an era of evaluation* (pp. 3–36). Palgrave Macmillan. doi:10.1007/978-3-030-75263-7_1

Franks, D., Dale, P., Hindmarsh, R., Fellows, C., Buckridge, M., & Cybinski, P. (2007). Interdisciplinary foundations: Reflecting on interdisciplinarity and three decades of teaching and research at Griffith University, Australia. *Studies in Higher Education*, 32 (2), 167–185. doi:10.1080/03075070701267228

Freeth, R., & Caniglia, G. (2020). Learning to collaborate while collaborating advancing interdisciplinary sustainability research. *Sustainability Science*, 15, 247–261. doi:10.1007/s11625-019-00701-z

Haklay, M.M. (2017). Beyond quantification: A role for citizen science and community science in a smart city. In R. Kitchin, T.P. Lauriault, & G. McArdle (Eds.), *Data and the city* (pp. 213–224). Routledge.

Horton, J., Macve, R., & Struyven, G. (2004). Qualitative research: Experiences in using semi-structured interviews. In C. Humphrey & B. Lee (Eds.), *The real life guide to accounting research* (pp. 339–357). Elsevier. doi:10.1016/B978-008043972-3/50022-0

Israilidis, J., Odusanya, K., & Mazhar, M.U. (2021). Exploring knowledge management perspectives in smart city research: A review and future research agenda. *International Journal of Information Management*, 56, 101989. doi:10.1016/j.ijinfomgt.2019.07.015

Kim, J. (2022). Smart city trends: A focus on five countries and 15 companies. *Cities*, 123, 103551. doi:10.1016/j.cities.2021.103551

Klein, J. (2005). Interdisciplinary teamwork: The dynamics of collaboration and integration. In S. Derry, C. Schunn, & M. Gernsbacher (Eds.), *Interdisciplinary collaboration: An emerging cognitive science* (pp. 23–59). Erlbaum.

Klein, J., & Newell, W. (1997). Advancing interdisciplinary studies. In J. Gaff & J. Ratcliff (Eds.), *Handbook of the undergraduate curriculum: A comprehensive guide to purposes, structures, practices and change* (pp. 393–415). Jossey-Bass.

Komiyama, H., & Takeuchi, K. (2006). Sustainability science: Building a new discipline. *Sustainability Science*, 1, 1–6.

Kuhn, T.S. (1970). *The structure of scientific revolutions* (2nd ed.). University of Chicago Press.

Lang, D.J., Wiek, A., Bergmann, M., Stauffacher, M., Martens, P., Moll, P., Swilling, M., & Thomas, C.J. (2012). Transdisciplinary research in sustainability science: Practice, principles, and challenges. *Sustainability Science*, 7, 25–43. doi:10.1007/s11625-011-0149-x

Lattuca, L. (2002). Learning interdisciplinarity. *Journal of Higher Education*, 73(6), 711–738.

Linköping University. (n.d.). *Strategic urban and regional planning*. https://liu.se/en/education/program/f7mur

Liu, X., Liang, L.L., & Liu, E. (Eds.). (2012). Science education research in China: Challenges and promises [Special issue]. *International Journal of Science Education*, 34(13), 1961–1970.

MacDonald, T.E., Javernick-Will, A., Austin-Breneman, J., Aranda, I., Salvinelli, C., Klees, R., Walters, J., Parmentier, M.J., Schaad, D., Shahi, A., Bedell, E., Platais, G., Brown, J., Gershenson, J., Watkins, D., Obonyo, E., Oyanedel-Craver, V., Olson, M., Lau, R., … Linden, K. (2022). Aligning learning objectives and approaches in global engineering graduate programs: Review and recommendations by an interdisciplinary working group. *Development Engineering*, 7, 100095. doi:10.1016/j.deveng.2022.100095

McNair, L., Davitt, M., & Batten, G.P. (2015). Outside the 'comfort zone': Impacts of interdisciplinary research collaboration on research, pedagogy, and disciplinary knowledge production. *Engineering Studies*, 7(1), 47–79. doi:10.1080/19378629.2015.1014817

Menken, S., & Keestra, M. (2020). *An introduction to interdisciplinary research*. Amsterdam University Press.

Mora, L., Deakin, M., Zhang, X., Batty, M., de Jong, M., Santi, P., & Appio, F.P. (2022). Assembling sustainable smart city transitions: An interdisciplinary theoretical perspective. *The Journal of Urban Technology*, 28(1–2), 1–27. doi:10.1080/10630732.2020.1834831

Newell, W. (2001). A reply to the respondents to 'A theory of interdisciplinary studies'. *Issues in Integrative Studies*, 19, 137–148. http://hdl.handle.net/10323/4384

Nissani, M. (1997). Ten cheers for interdisciplinarity: The case for interdisciplinary knowledge and research. *The Social Science Journal*, 34(2), 201–216. doi:10.1016/S0362-3319(97)90051-3

Norwegian University of Science and Technology. (n.d.). *NTNU smart and sustainable cities*. https://www.ntnu.edu/smartcities

Pautasso, M., & Pautasso, C. (2010). Peer reviewing interdisciplinary papers. *European Review*, 18(2), 227–237. doi:10.1017/S1062798709990275

Pavalko, R.M. (1988). *Sociology of occupations and professions*. F.E. Peacock.

Petrie, H.G. (1976). Do you see what I see? The epistemology of interdisciplinary inquiry. *Educational Researcher*, 5(10), 9–15. doi:10.3102/0013189X005C02009

Petts, J., Owens, S., & Bulkeley, H. (2008). Crossing boundaries: Interdisciplinarity in the context of urban environments. *Geoforum*, 39(2), 593–601.

Reid, W.V., Chen, D., Goldfarb, L., Hackmann, H., Lee, Y.T., Mokhele, K., Ostrom, E., Raivio, K., Rockström, J., Schellnhuber, H.J., & Whyte, A. (2010). Earth system science for global sustainability: Grand challenges. *Science*, 330(6006), 916–917. doi:10.1126/science.1196263

Rhoten, D., & Parker, A. (2004). Risks and rewards of an interdisciplinary research path. *Science*, 306(5704), 2046. doi:10.1126/science.1103628

Riggio, R.E. (2011). Is leadership studies a discipline? In M. Harvey & R.E. Riggio (Eds.), *Leadership studies: The dialogue of disciplines* (pp. 9–19). Edward Elgar.

Rylance, R. (2015). Grant giving: Global funders to focus on interdisciplinarity. *Nature*, 525, 313–315. doi:10.1038/525313a

Schuitema, G., & Sintov, N.D. (2017). Should we quit our jobs? Challenges, barriers and recommendations for interdisciplinary energy research. *Energy Policy*, 101, 246–250. doi:10.1016/j.enpol.2016.11.043

Shell, R., Soe, R.M., Wang, R., & Rassõlkin, A. (2021). Autonomous vehicle shuttle in smart city testbed. In C. Zachäus & G. Meyer (Eds.), *Intelligent system solutions for auto mobility and beyond* (pp. 143–157). Springer. doi:10.1007/978-3-030-65871-7_11

Staupe-Delgado, R., Abdel-Fattah, D., & Pursiainen, C. (2022). A discipline without a name? Contrasting three fields dealing with hazards and disaster. *International Journal of Disaster Risk Reduction*, 70, 102751.

Szostak, R. (2002). How to do interdisciplinarity: Integrating the debate. *Issues in Integrative Studies*, 20, 103–122. http://hdl.handle.net/10323/4393

Thompson, M.A., Owen, S., Lindsay, J.M., Leonard, G.S., & Cronin, S.J. (2017). Scientist and stakeholder perspectives of transdisciplinary research: Early attitudes, expectations, and tensions. *Environmental Science & Policy*, 74, 30–39. doi:10.1016/j.envsci.2017.04.006

United Nations. (2015). *Transforming our world: The 2030 agenda for sustainable development*. https://sdgs.un.org/2030agenda

University of Stavanger. (2024). *Smart city research*. https://www.uis.no/en/research/smart-city-research

van Noorden, R. (2015). Interdisciplinary research by the numbers. *Nature*, 525, 306–307. doi:10.1038/525306a

van Rijnsoever, F.J., & Hessels, L.K. (2011). Factors associated with disciplinary and interdisciplinary research collaboration. *Research Policy*, 40, 463–472.

Wang, B., Wu, C., Huang, L., Kang, L., & Lei, Y. (2020). Safety science as a new discipline in China. *Safety Science*, 121, 201–214.

Washburn, D., & Sindhu, U. (2010). *Helping CIOs understand 'smart city' initiatives*. Forrester Research. http://www.uwforum.org/upload/board/forrester_help_cios_smart_city.pdf

Weichselgartner, J., & Truffer, B. (2015). From knowledge co-production to transdisciplinary research: Lessons from the quest to produce socially robust knowledge. In B. Werlen (Ed.), *Global sustainability: Cultural perspectives and challenges for transdisciplinary integrated research* (pp. 89–106). doi:10.1007/978-3-319-16477-9_5

Zygiaris, S. (2013). Smart city reference model: Assisting planners to conceptualize the building of smart city innovation ecosystems. *Journal of the Knowledge Economy*, 4(2), 217–231. doi:10.1007/s13132-012-0089-4

4 Transformative agency in urban experimentation
The role of intermediaries and boundary spanners

Veronika Lorentzen and Oluf Langhelle

Introduction

In recent discourse, numerous categories of cities have emerged, each prefixed with descriptors such as 'sustainable', 'green', 'smart', and 'resilient' (de Jong et al., 2015). Central to these concepts is the principle of urban experimentation, which offers a framework for transforming ambitious urban visions into concrete actions that promote more liveable and sustainable futures (Sengers et al., 2016).

Urban experimentation represents a convergence of diverse urban aspirations, from developing low-carbon cities to supporting grassroot movements advocating social justice and sustainable living (Ansell & Bartenberger, 2016; Kivimaa et al., 2017; Torrens et al., 2019; Turnheim et al., 2018). It reflects a recognition of experimentation as essential for navigating the complex challenges of modern urban landscapes, acting as a bridge between various urban development goals (Bulkeley, 2023; Hodson et al., 2018; Sengers et al., 2019; Torrens & von Wirth, 2021; Voytenko et al., 2016).

Within this evolving landscape, intermediaries have emerged as key agents in the urban realm (Barnes, 2017; Moss, 2009; Torrens & von Wirth, 2021; van Lente et al., 2003). These entities, both formal organizations and individual actors, act as connectors, facilitating the transfer of knowledge and practices necessary for iterative innovation in urban experimentation. Their role extends beyond knowledge transfer, influencing the design, implementation, and scaling of experiments.

While research has explored intermediaries' roles in amplifying innovations and sharing knowledge (Ehnert, 2023; Matschoss & Heiskanen, 2017), the dynamics of micro-level agency – highlighting individual contributions and interactions within small groups – remain underexamined (Smedlund, 2006). This highlights the need to examine the role of agency in urban sustainability transitions, focusing on how individual and collective actions impact experimental endeavours (Bæk-kelund, 2021; Beer et al., 2023; Ehnert, 2023; Fuenfschilling et al., 2019; Hansen & Coenen, 2015; Isaksen et al., 2019; Rekers & Stihl, 2021; Roebke et al., 2022). These actions can aim to drive change or preserve existing structures (Jolly et al., 2020; Roebke et al., 2022). The drive for change is often referred to as *transformative* or *change agency*, while efforts to maintain are known as *maintenance* or *reproductive agency* (Coe & Jordhus-Lier, 2011; Roebke et al., 2022).

DOI: 10.4324/9781003498650-7

Despite the potential significance, understanding of boundary spanners' roles, functions, and effects on urban experimentation remains limited (Torrens et al., 2019). There's a need to examine their practices to understand their agency and its impact on socio-technical systems (Avelino & Wittmayer, 2017; Ehnert, 2023; Schot et al., 2016; Westley et al., 2013; Wittmayer et al., 2017). By focusing on agency, we can link the actions and strategies of various actors to outcomes, incorporating a temporal dimension to the interplay of structure and agency in urban experimentation (Archer, 1982; Archer et al., 2013; Roebke et al., 2022).

This chapter aims to investigate the roles of boundary spanners in sustainability transitions within urban experimental projects and identify critical aspects of boundary spanning. Our goal is to contribute to the discussion on agency in urban transitions (Kivimaa et al., 2019; Markard et al., 2012; Smink et al., 2015), offering insights for local governments to better support intermediaries and boundary spanners (Roebke et al., 2022).

Context and method

Norwegian cities vary in their approach to urban experimentation. While Stavanger initially focused on innovation through technological advancements in its smart city initiatives (Stavanger Kommune, n.d.), recent efforts have increasingly incorporated citizen involvement, as discussed in other chapters of this volume. Conversely, Trondheim has consistently emphasized co-creation and placing citizens at the centre of development (Trondheim Kommune, n.d.). For researching the role of agency in sustainability transitions, Trondheim's approach is particularly relevant due to its emphasis on stakeholder engagement and collaborative practices.

Trondheim has been at the forefront of urban experimentation, notably with its long-standing project Svartlamon (Sager, 2011, 2018, 2024). Originating from grassroots actions by residents and formalized by city council decisions in 2001 and 2006 (Trondheim Kommune, 2001, 2006), Svartlamon serves as an urban ecological experimental area. As Stavanger now aims to develop experimental areas such as living labs, the experiences from Trondheim may provide valuable insights.

Svartlamon covers 32.1 acres predominantly owned by the municipality. It is managed by the Svartlamon Housing Foundation and the Svartlamon Culture and Business Foundation, which act as intermediaries between the municipality and the residents. Initially designated for industrial use, the area transformed when young people occupied vacant, poorly maintained houses, leading to its re-designation as a residential area in 1998. Svartlamon's objective is to foster innovative ideas for sustainable housing and business development, incorporating experimental approaches to housing, technology, architecture, planning, and collaboration.

We employed a qualitative case study approach to examine the roles of boundary spanners in Svartlamon, consistent with the methodology used in other studies (van Meerkerk & Edelenbos, 2018). Data were collected through semi-structured interviews with key stakeholders, including boundary spanners, municipality officials, residents, business figures, and researchers involved with Svartlamon (R1–37; see appendix). The interviews were conducted as part of the broader PhD project of the first author and are used selectively in this chapter. Additional data came from observations and reviews of project documents and archival records. Thematic analysis was conducted using NVivo software (QSR International, 2024) to process and analyze the data. A full list of interviews is provided in the appendix.

Theoretical and empirical synthesis

Intermediaries play key roles in sustainability transitions by facilitating knowledge gathering, learning, networking, brokering, managing innovation processes, visioning, and legitimizing institutional change (Kivimaa, 2014; Sovacool et al., 2020). These roles align with functions critical for establishing new niches (Schot & Geels, 2008; van der Laak et al., 2007) and supporting technological innovation systems that foster transitions (Suurs & Hekkert, 2009). However, these connections have not been explicitly addressed in the literature (Kivimaa, 2014).

Intermediaries often appear as organizations operating between communities, governments, and the private sector, coordinating collaborative efforts towards common goals. In Svartlamon, these roles are exemplified by two intermediary organizations outlined in the regulation plans of 2001 and 2006: the Housing Foundation and the Culture and Business Foundation (Trondheim Kommune, 2001, 2006). These entities are designed to guide the Svartlamon project towards sustainable urban development, manage residential and commercial spaces, and act as bridges between the municipality and the residents' community.

The Housing Foundation manages 35 houses with a dedicated team, mediating between residents and the Trondheim municipality. Its governance combines resident-elected members and political appointees, striving for consensus and participatory decision making.

The Culture and Business Foundation fosters an environment for artistic expression and entrepreneurship. It supports artistic experimentation and cultural enterprises, emphasizing affordable access for young creatives and non-commercial ventures.

Both foundations embody the theoretical functions of intermediaries within the Svartlamon experiment, mediating among stakeholders and promoting a sustainable, creative, and participatory urban community. Their efforts demonstrate a practical application of intermediary roles in real-world sustainability transitions.

Translating these organizational roles into tangible impacts often depends on the individuals within these entities – the boundary spanners. They are conceptualized as agents navigating boundaries between different domains, facilitating communication, cooperation, and the exchange of resources and information (van Meerkerk & Edelenbos, 2018).

The discourse on boundary spanners unfolds through multiple lenses, with a notable emphasis on their roles, activities, and functions. This research adopts the action perspective, focusing on the agency's role in understanding boundary-spanning behaviour (van Meerkerk & Edelenbos, 2018). According to this perspective, boundary spanners are characterized by their behaviours and activities rather than formal roles. Key activities include building relationships, coordinating internal and external linkages, sharing information and knowledge, and entrepreneurial efforts.

The action perspective categorizes these activities into four principal roles:

- **The Fixer**: solves problems in cross-boundary connections by aligning an organization's practices with external partners
- **The Bridger**: creates connections between parties and promotes cross-boundary initiatives
- **The Broker**: facilitates and mediates interactions between actors with different objectives
- **The Innovative Entrepreneur**: explores new ideas and seeks opportunities to develop and support initiatives

Analysis: boundary spanning roles in Svartlamon

Drawing from these theoretical insights, this study identifies three variations of boundary spanners (BSs) among the representatives of the Svartlamon foundations through analysis of their activities.

Boundary spanner as the Fixer and the Broker

BS1's key activities were identified by a representative of the Housing Foundation who enacted the Fixer and Broker roles described by van Meerkerk and Edelenbos (2018).

As a *Fixer*, BS1 scans external information, translating complex policies and regulations for residents unfamiliar with bureaucratic processes. A resident explained:

> BS1 is there to inform us about legal matters, for example, if we build too close to the train tracks. There are so many rules. BS1 knows all about those laws and applications – everything related to Norwegian bureaucracy. (R9)

BS1 highlighted that boundary spanning involves finding common ground between residents and the municipality while overcoming barriers:

> It's a tricky balance. The municipality has strict processes and long response times, but residents' initiatives are often immediate. Sometimes, I feel my role is to just say, 'I don't want to hear about it … just do it, and we'll see'. (R1)

BS1 supports residents' initiatives with minimal interference:

> I try not to intervene. I believe the more they handle themselves, the better. (R1)

In addition to these activities, BS1 focused on gathering knowledge about actors in the municipality and politicians supportive of Svartlamon:

> I work hard to find the right people. Once I find them, I stick with them because they help me navigate. (R1)

In the *Broker* role, BS1 aligns residents' interests with broader project objectives, guiding what is feasible within municipal constraints:

> BS1 informs us and shares their opinion, saying things like, 'This won't happen because the fire department needs this space, so forget about it'. (R9)

BS1 extended the role by forging partnerships to create public value, collaborating with the Trondheim Housing Foundation to transform vacant housing into affordable options:

> I went to a board meeting. They own about 900 apartments in Trondheim. I told them about Svartlamon and said, 'Just give me a block, and I'll start the third housing sector in Norway'. (R1)

BS1's roles as a Fixer and Broker allow them to significantly shape the trajectory of Svartlamon's experimentation. By leveraging the broad definitions of the project, BS1 actively steers its direction:

> It's fun because, with a broad concept, you can define it yourself, as long as you stay within the framework. (R1)

BS1's engagement in political discourse and collaboration with key stakeholders further impacts the directionality of urban transitions. They successfully advocated for key changes in the housing policy:

> We got the municipality to include working with non-profit foundations in the housing policy plan, helping to develop the third housing sector. (R1)

BS1 also actively builds niches through creative outreach. By introducing ideas and supporting others, they help foster new initiatives:

> I throw out ideas, hoping someone with more capacity will say, 'That's a good idea'. (R1)

BS1 also empowers niches to grow:

> When someone wanted to occupy houses, I provided what they needed to start. I'm like a resource bank. (R1)

The role of boundary spanners like BS1 in urban transitions comes with significant challenges. Balancing experimentation with municipal bureaucracy creates tension:

> The goal is to try new things, but meeting the municipality's hierarchy often leads to a crash. (R1)

BS1 acknowledges the ongoing nature of these challenges:

> If you solve one problem, another will arise. It's a complex process you have to accept. (R1)

Despite this, their mindset remains solution-oriented:

> When I get a problem, my first thought is always, 'How can we make this happen?' (R1)

These insights highlight the complex dynamics of boundary spanning in urban experiments. BS1 plays a crucial role in facilitating transitions while navigating personal challenges and stresses.

Boundary spanner as the Bridger and Innovative Entrepreneur

BS2's activities by a representative from the Culture and Business Foundation align with the roles of a Bridger and an Innovative Entrepreneur (van Meerkerk & Edelenbos, 2018). As a long-time resident of Svartlamon, BS2 has earned substantial trust from other residents, enabling them to bridge between the municipality, business actors, and residents engaged in culture and arts.

As a *Bridger*, BS2 has helped position Svartlamon within the city's cultural and national scene through ambassadorial activities:

> I'm managing the venue, which invites people from Trondheim and across Norway to visit, attend concerts, and explore the area. It has opened up Svartlamon to many. (R12)

BS2 has played a key role in securing resources and funding for cultural activities and artist support through the municipality's grant scheme. They highlighted the synergy between residents' and business activities, a core aspect of the Bridger role.

BS2 has also been involved in negotiating with local politicians to secure support for Svartlamon:

> We are in discussions with politicians, but it's long-term work. We're talking to all the different parties. (R12)

As an *Innovative Entrepreneur*, BS2 drives initiatives by adjusting formal structures to support experimental artistic expressions. However, economic challenges limit this:

> We'd like to be more experimental, but it's an economic issue. We pay rent to
> the municipality, so we have to reduce time spent on experimentation to make
> money. (R12)

To address these constraints, BS2 is focused on securing the future of the Culture
and Business Foundation through new partnerships, including collaborations with
a local pub.

As an Innovative Entrepreneur, BS2 has explored expanding the Svartlamon
concept and regulatory model to broader scales, advocating for a value system that
embraces experimentation and risk-taking:

> With EU regulations, society is becoming more streamlined. I'm not saying
> we shouldn't follow them, but we need space to experiment – maybe some-
> thing good will come out of it. (R12)

BS2 envisions Svartlamon as a model for other Norwegian cities to enable more
experimentation:

> If this would be a success, you could make similar regulations that allow more
> experimentation. (R12)

While BS1 and BS2 embodied traditional boundary spanning roles, the landscape
is evolving. With BS1 leaving to pursue other ambitions, BS3 stepped in, bringing
a new perspective and suggesting new boundary spanning profiles within Svartlamon's ongoing experiment.

Boundary spanner as a 'Sustainer' and a 'Shielder'

BS3's role within the Svartlamon community extends beyond traditional boundary-spanning roles described by van Meerkerk and Edelenbos (2018).

BS3 reflects on this (emerging) role by relating to the personal experiences:

> In some way, I view myself as in a theatre. There are the accountants and the
> people who don't work with creativity – that's kind of my job. I maintain the
> structure and system so it's possible to live here, be creative, and experiment.
> The residents are the ones leading the experiment; they have to be part of it
> and drive it forward. (R28)

Based on the quotation and the concept of maintaining the structure to enable
others' creativity and experimentation, a fitting role for BS3 might be the 'Sustainer'. This term reflects someone who maintains the foundation and ensures stability, allowing the creative and experimental efforts of the residents to thrive. The
Sustainer is essential in any innovative environment, providing the consistency and
support needed for experiments to be conducted without concerns about the
structural integrity of the platform.

This role seems to be a newly emerging aspect of boundary spanning in urban experimental contexts, highlighting the evolution of these roles in response to the dynamic needs of sustainability transitions. BS3's contributions emphasize the importance of stability and support in fostering an environment where experimentation and a community's creative potential can flourish.

Moreover, it seems that another emerging role is beginning to take shape through the narrative shared by BS3:

> One of the most bureaucratically impossible projects is those two buses. ... There's a couple living in one of them. The police tried to evict them, considering it living in a tent, which isn't allowed in the city area. They came to remove them but didn't succeed. We had a lawyer who works pro bono for us. So when the police came, they looked down on these people living in a bus. But then me and my lawyer showed up formally dressed, defending them. The police were completely appalled to meet a professional lawyer and a bureaucrat protecting this bus. I think it's the most rewarding part of my job – to represent something that many people don't think is represented, to build a system that nobody thinks would exist. (R25)

BS3 seems to be taking on a role that can be proposed as a 'Shielder' in the context of niche shielding within sustainability transitions. This role involves protecting innovative and experimental practices from external bureaucratic and legal pressures. The Shielder acts as a guardian, ensuring that creative and alternative lifestyle practices can continue without being dismantled by conventional enforcement.

BS3's defense of residents living in buses exemplifies this protective role. By utilizing formal knowledge and resources, like legal assistance, BS3 safeguards the community's niche experiments. This action ensures the persistence and resilience of the creative ecosystem against external forces that may not understand or value the community's experimental approach. It is a strategic role that navigates the broader system's rules to create a buffer for the niche, allowing it to develop in a supportive environment.

BS3 also faces role stressors, dealing with external pressures like the potential demolition due to train track expansion, which has a personal impact:

> The residents are trying to organize protests and movements to stop it. It's kind of sad ... very sad. I was depressed for a week after that. (R17)

By having both a Sustainer and a Shielder, Svartlamon benefits from roles that foster growth and expansion while providing robust defence. These roles ensure that innovative practices are not prematurely stifled by regulatory, legal, or social challenges.

Discussion and conclusion

The Svartlamon case illustrates how intermediary organizations, such as the Housing Foundation and the Culture and Business Foundation, facilitate urban sustainability transitions. These entities bridge communities and stakeholders, acting as connectors that support sustainable urban living and cultural vibrancy. Their roles align with theoretical intermediary functions, including knowledge exchange, networking, and innovation management (Kivimaa, 2014; Sovacool et al., 2020).

Examining the activities of boundary spanners within these organizations reveals how individual actions materialize the broader functions of intermediaries (van Meerkerk & Edelenbos, 2018). BS1 and BS2 fulfilled traditional boundary-spanning roles – BS1 as a Fixer and Broker and BS2 as a Bridger and Innovative Entrepreneur (van Meerkerk & Edelenbos, 2018). Their efforts demonstrate the importance of personal agency in driving change and shaping the direction of urban transitions, aligning with the concept of transformative or change agency, where actions aim to initiate and realize change (Coe & Jordhus-Lier, 2011; Roebke et al., 2022).

BS1 navigated bureaucratic challenges and aligned residents' initiatives with municipal regulations, while BS2 expanded the project's cultural impact and secured resources. These roles were effective during the earlier stages of Svartlamon's development, suggesting that transformative agency is essential in initiating and guiding early experimentation.

The emergence of new roles with BS3, who joined in the later stages of the project, indicates an evolution in boundary-spanning needs as the project matures. BS3 embodied the roles of 'Sustainer' and 'Shielder', focusing on maintaining foundational structures that enable ongoing experimentation and protecting the community from external pressures. This reflects maintenance or reproductive agency, where actions aim to preserve and stabilize existing structures (Coe & Jordhus-Lier, 2011; Jolly et al., 2020; Roebke et al., 2022). BS3's efforts were essential in ensuring the longevity and resilience of the experimental initiatives, highlighting the significance of maintenance agency in the sustainability transition process.

The case of Svartlamon demonstrates that outcomes in change processes result from a combination of transformative and maintenance agency (Bækkelund, 2021; Jolly et al., 2020; Roebke et al., 2022). The change-oriented boundary-spanning actions by BS1 and BS2 set the stage for innovation and experimentation, while BS3's maintenance-oriented roles ensured the sustainability and continued development of those initiatives. The interplay between these types of agency underscores the importance of both catalyzing change and sustaining it over time.

Importantly, role stressors and emotional challenges were present in both types of agency, affecting boundary spanners engaged in driving change as well as those focused on maintaining structures. An interesting observation is that role stressors and emotional challenges were reported by BS1 and BS3 but not by BS2. BS2, being a resident and insider of Svartlamon, may have developed resilience to these stressors. This contrast implies that insiders might experience less role strain due to their deeper integration within the community, a notion that merits further research to understand how insider status influences the well-being of boundary spanners.

These insights have practical implications. Recognizing the changing nature of boundary-spanning roles throughout a project's lifecycle highlights the need for tailored support systems that address specific challenges faced by boundary spanners. Providing resources and frameworks to mitigate potential burnout and promote sustainable engagement can enhance the effectiveness of these key agents of change.

Future research could explore how boundary spanners' roles evolve as urban experiments mature and how the balance between change and maintenance agency influences the speed and trajectory of sustainability transitions. Investigating factors that contribute to resilience among boundary spanners, particularly the impact of being an insider versus an outsider, could offer valuable guidance for supporting their well-being.

The Svartlamon case serves as a lens to examine the interplay between structure and agency in sustainability transitions. Understanding the diverse roles of boundary spanners enriches both theoretical perspectives and practical applications in urban sustainability. As cities worldwide pursue smart city initiatives, incorporating lessons from Svartlamon may inform strategies that balance technological innovation with community engagement, ultimately contributing to more inclusive and resilient urban futures.

Appendix

Table A.1 Interview list

Code	Interviewee	Date	Location
R1	Representative from Housing Foundation, potential BS1	November 2, 2020	Trondheim
R2	Representative from Bopilot project	November 9, 2020	Online
R3	Representative from CityExchange, UN Centre of excellence	December 4, 2020	Online
R4	Representative from academia	November 25, 2020	Online
R5	Representative from Bopilot project	December 9, 2020	Online
R6	Representative from Trondheim municipality	November 5, 2020	Trondheim
R7	Network representative	November 18, 2020	Online
R8	Representative from academia	November 18, 2020	Online
R9	Resident, Svartlamon	November 12, 2020	Online
R10	Resident, Svartlamon	November 10, 2020	Online
R11	Resident, Svartlamon	November 10, 2020	Online
R12	Culture and Business Foundation representative, potential BS2	February 22, 2021	Online

Code	Interviewee	Date	Location
R13	Representative from non-profit organization in Svartlamon	July 9, 2023	Trondheim
R14	Representative from non-profit organization in Svartlamon	July 9, 2023	Trondheim
R15	Resident, Svartlamon	July 11, 2023	Trondheim
R16	Business owner in Svartlamon	July 11, 2023	Trondheim
R17	Representative from Housing Foundation, potential BS3	July 12, 2023	Trondheim
R18	Self-builder in Svartlamon	July 6, 2023	Online
R19	Self-builder in Svartlamon	July 6, 2023	Online
R20	Non-profit organization representative	July 12, 2023	Trondheim
R21	Resident, Svartlamon	July 15, 2023	Trondheim
R22	Postman in Svartlamon	July 15, 2023	Trondheim
R23	Kindergarten and Remida centre representative	July 13, 2023	Trondheim
R24	Business owner in Svartlamon	July 8, 2023	Trondheim
R25	Representative from Housing Foundation, potential BS3	July 12, 2023	Trondheim
R26	Gallery volunteer/administrator	July 12, 2023	Trondheim
R27	Business representative	July 13, 2023	Trondheim
R28	Walk-and-talk with the representative from Housing Foundation, potential BS3	July 13, 2023	Trondheim
R29	Walk-and-talk with Mellamon	July 18, 2023	Trondheim
R30	Self-builder	November 26, 2019	Online
R31	Self-builder	November 26, 2019	Online
R32	Visitor of Svartlamon	July 9, 2023	Trondheim
R33	Neighbourhood resident	July 10, 2023	Trondheim
R34	Walk-and-talk with self-builders	May 29, 2019	Trondheim
R35	Researcher in the sector	October 27, 2020	Stavanger
R36	Resident, Svartlamon	August 10, 2023	Online
R37	Resident, Svartlamon	August 24, 2023	Online

References

Ansell, C.K., & Bartenberger, M. (2016). Varieties of experimentalism. *Ecological Economics*, 130, 64–73. doi:10.1016/j.ecolecon.2016.05.016

Archer, M.S. (1982). Morphogenesis versus structuration: On combining structure and action. *The British Journal of Sociology*, 33(4), 455–483. doi:10.2307/589357

Archer, M.S., Bhaskar, R., Collier, A., Lawson, T., & Norrie, A. (2013). *Critical realism: Essential readings*. Routledge.

Avelino, F., & Wittmayer, J. (2017). Interlude: A multi-actor perspective on urban sustainability transitions. In N. Frantzeskaki, K. Hölscher, M. Bach, & F. Avelino (Eds.), *Urban sustainability transitions* (pp. 272–284). Routledge.

Bækkelund, N.G. (2021). Change agency and reproductive agency in the course of industrial path evolution. *Regional Studies*, 55(4), 757–768. doi:10.1080/00343404.2021.1893291

Barnes, J. (2017). *User-intermediaries and the local embedding of low carbon technologies.* SSRN. doi:10.2139/ssrn.3019957

Beer, A., Barnes, T., & Horne, S. (2023). Place-based industrial strategy and economic trajectory: Advancing agency-based approaches. *Regional Studies*, 57(6), 984–997. doi:10.1080/00343404.2021.1947485

Bulkeley, H. (2023). The condition of urban climate experimentation. *Sustainability: Science, Practice and Policy*, 19(1), Article2188726. doi:10.1080/15487733.2023.2188726

Coe, N.M., & Jordhus-Lier, D.C. (2011). Constrained agency? Re-evaluating the geographies of labour. *Progress in Human Geography*, 35(2), 211–233. doi:10.1177/0309132510366746

de Jong, M., Joss, S., Schraven, D., Zhan, C., & Weijnen, M. (2015). Sustainable–smart–resilient–low carbon–eco–knowledge cities: Making sense of a multitude of concepts promoting sustainable urbanization. *Journal of Cleaner Production*, 109, 25–38. doi:10.1016/j.jclepro.2015.02.004

Ehnert, F. (2023). Review of research into urban experimentation in the fields of sustainability transitions and environmental governance. *European Planning Studies*, 31(1), 76–102. doi:10.1080/09654313.2022.2070424

Fuenfschilling, L., Frantzeskaki, N., & Coenen, L. (2019). Urban experimentation & sustainability transitions. *European Planning Studies*, 27(2), 219–228. doi:10.1080/09654313.2018.1532977

Hansen, T., & Coenen, L. (2015). The geography of sustainability transitions: Review, synthesis and reflections on an emergent research field. *Environmental Innovation and Societal Transitions*, 17, 92–109. doi:10.1016/j.eist.2014.11.001

Hodson, M., Evans, J., & Schliwa, G. (2018). Conditioning experimentation: The struggle for place-based discretion in shaping urban infrastructures. *Environment and Planning C: Politics and Space*, 36(8), 1480–1498. doi:10.1177/2399654418784308

Isaksen, A., Jakobsen, S.-E., Njøs, R., & Normann, R. (2019). Regional industrial restructuring resulting from individual and system agency. *Innovation: The European Journal of Social Science Research*, 32(1), 48–65. doi:10.1080/13511610.2018.1496322

Jolly, S., Grillitsch, M., & Hansen, T. (2020). Agency and actors in regional industrial path development: A framework and longitudinal analysis. *Geoforum*, 111, 176–188. doi:10.1016/j.geoforum.2020.02.013

Kivimaa, P. (2014). Government-affiliated intermediary organisations as actors in system-level transitions. *Research Policy*, 43(8), 1370–1380. doi:10.1016/j.respol.2014.02.007

Kivimaa, P., Boon, W., Hyysalo, S., & Klerkx, L. (2019). Towards a typology of intermediaries in sustainability transitions: A systematic review and a research agenda. *Research Policy*, 48(4), 1062–1075. doi:10.1016/j.respol.2018.10.006

Kivimaa, P., Hildén, M., Huitema, D., Jordan, A., & Newig, J. (2017). Experiments in climate governance: A systematic review of research on energy and built environment transitions. *Journal of Cleaner Production*, 169, 17–29. doi:10.1016/j.jclepro.2017.01.027

Markard, J., Raven, R., & Truffer, B. (2012). Sustainability transitions: An emerging field of research and its prospects. *Research Policy*, 41(6), 955–967. doi:10.1016/j.respol.2012.02.013

Matschoss, K., & Heiskanen, E. (2017). Making it experimental in several ways: The work of intermediaries in raising the ambition level in local climate initiatives. *Journal of Cleaner Production*, 169, 85–93. doi:10.1016/j.jclepro.2017.03.037

Moss, T. (2009). Intermediaries and the governance of sociotechnical networks in transition. *Environment and Planning A: Economy and Space*, 41(6), 1480–1495. doi:10.1068/a4116

QSR International. (2024). *NVivo* (Version 15) [Computer software]. https://www.qsrinternational.com/nvivo-qualitative-data-analysis-software/home

Rekers, J.V., & Stihl, L. (2021). One crisis, one region, two municipalities: The geography of institutions and change agency in regional development paths. *Geoforum*, 124, 89–98. doi:10.1016/j.geoforum.2021.05.012

Roebke, L., Grillitsch, M., & Coenen, L. (2022). Assessing change agency in urban experiments for sustainability transitions. *Environmental Innovation and Societal Transitions*, 45, 214–227. doi:10.1016/j.eist.2022.10.007

Sager, T. (2011). Activist planning: A response to the widespread marginalization of radical planning ideas. *Progress in Planning*, 76(1), 1–60. doi:10.1016/j.progress.2011.09.001

Sager, T. (2018). Planning by intentional communities: An understudied form of activist planning. *International Journal of Urban and Regional Research*, 42(4), 684–695. doi:10.1111/1468-2427.12655

Sager, T. (2024). Grassroots urban labs: Experimentation and radical planning in alternative microsocieties. *Nordic Journal of Urban Studies*, 4(1), 1–26. doi:10.18261/njus.4.1.3

Schot, J., & Geels, F.W. (2008). Strategic niche management and sustainable innovation journeys: Theory, findings, research agenda, and policy. *Technology Analysis & Strategic Management*, 20(5), 537–554. doi:10.1080/09537320802292651

Schot, J., Kanger, L., & Verbong, G. (2016). The roles of users in shaping transitions to new energy systems. *Nature Energy*, 1(5), Article16054. doi:10.1038/nenergy.2016.54

Sengers, F., Berkhout, F., Wieczorek, A.J., & Raven, R. (2016). Unpacking notions of experimentation for sustainability. In J. Evans, A. Karvonen, & R. Raven (Eds.), *The experimental city* (pp. 15–31). Routledge.

Sengers, F., Wieczorek, A.J., & Raven, R. (2019). Experimenting for sustainability transitions: A systematic literature review. *Technological Forecasting and Social Change*, 145, 153–164. doi:10.1016/j.techfore.2016.08.031

Smedlund, A. (2006). The roles of intermediaries in a regional knowledge system. *Journal of Intellectual Capital*, 7(2), 204–220. doi:10.1108/14691930610661863

Smink, M., Negro, S.O., Niesten, E., & Hekkert, M.P. (2015). How mismatching institutional logics hinder niche–regime interaction and how boundary spanners intervene. *Technological Forecasting and Social Change*, 100, 225–237. doi:10.1016/j.techfore.2015.07.004

Sovacool, B.K., Turnheim, B., Martiskainen, M., Brown, D., & Kivimaa, P. (2020). Guides or gatekeepers? Incumbent-oriented transition intermediaries in a low-carbon era. *Energy Research & Social Science*, 66, Article101490. doi:10.1016/j.erss.2020.101490

Stavanger Kommune. (n.d.). *Smartbyen Stavanger*. https://www.stavanger.kommune.no/samfunnsutvikling/smartbyen-stavanger/

Suurs, R.A.A., & Hekkert, M.P. (2009). Cumulative causation in the formation of a technological innovation system: The case of biofuels in the Netherlands. *Technological Forecasting and Social Change*, 76(8), 1003–1020. doi:10.1016/j.techfore.2009.03.002

Torrens, J., Schot, J., Raven, R., & Johnstone, P. (2019). Seedbeds, harbours, and battlegrounds: On the origins of favourable environments for urban experimentation with sustainability. *Environmental Innovation and Societal Transitions*, 31, 211–232. doi:10.1016/j.eist.2018.11.003

Torrens, J., & von Wirth, T. (2021). Experimentation or projectification of urban change? A critical appraisal and three steps forward. *Urban Transformations*, 3(1), Article8. doi:10.1186/s42854-021-00025-1

Trondheim Kommune. (n.d.). *Hva betyr smartby på Trondheim?* https://www.trondheim.komm une.no/aktuelt/om-kommunen/bk/barekraft/smartby/hva-betyr-smartby-pa-trondheim/

Trondheim Kommune. (2001). *Reguleringsplan med bestemmelser for Svartlamoen (0149/01)*. http://svartlamon.org/wp-content/uploads/2013/10/1176894434_Bystyrevedtak_ om_Svartlamon.doc

Trondheim Kommune. (2006). *Bestemmelser til endret reguleringsplan for Svartlamoen (R 219b)*. http://kart5.nois.no/trondheimbraarkiv/getfile.aspx?id=20133810

Turnheim, B., Kivimaa, P., & Berkhout, F. (2018). Beyond experiments: Innovation in climate governance. In B. Turnheim, P. Kivimaa, & F. Berkhout (Eds.), *Innovating climate governance* (pp. 1–26). Cambridge University Press. doi:10.1017/9781108277679.002

van der Laak, W.W.M., Raven, R.P.J.M., & Verbong, G.P.J. (2007). Strategic niche management for biofuels: Analysing past experiments for developing new biofuel policies. *Energy Policy*, 35(6), 3213–3225. doi:10.1016/j.enpol.2006.11.009

van Lente, H., Hekkert, M., Smits, R., & van Waveren, B. (2003). Roles of systemic intermediaries in transition processes. *International Journal of Innovation Management*, 7(3), 247–279. doi:10.1142/S1363919603000817

van Meerkerk, I., & Edelenbos, J. (2018). *Boundary spanners in public management and governance: An interdisciplinary assessment*. Edward Elgar.

Voytenko, Y., McCormick, K., Evans, J., & Schliwa, G. (2016). Urban living labs for sustainability and low carbon cities in Europe: Towards a research agenda. *Journal of Cleaner Production*, 123, 45–54. doi:10.1016/j.jclepro.2015.08.053

Westley, F.R., Tjornbo, O., Schultz, L., Olsson, P., Folke, C., Crona, B., & Bodin, Ö. (2013). A theory of transformative agency in linked social–ecological systems. *Ecology and Society*, 18(3), Article27. doi:10.5751/ES-05072-180327

Wittmayer, J.M., Avelino, F., van Steenbergen, F., & Loorbach, D. (2017). Actor roles in transition: Insights from sociological perspectives. *Environmental Innovation and Societal Transitions*, 24, 45–56. doi:10.1016/j.eist.2016.10.003

5 A pragmatist approach to the smart city concept and practice

Kristiane M.F. Lindland

Introduction

Cities can be understood as endless improvement projects under continuous development and transformation. The smart city concept was initially a concept for urban development connected to how digital solutions can contribute to realize better urban solutions, whether it is connected to mobility, technical and physical infrastructure, energy solutions, or safety and security (Zheng et al., 2020). A central aim of smart city research and innovation is thus to develop, explore, and make use of new technologies, business models, methods, and organizational practices that can work across disciplinary challenges and geographical contexts. From addressing possible technical and digital solutions to urban challenges, smart city research on solutions has increasingly been drawing more on social sciences, among other academic fields (Echebarria et al., 2021; Zheng et al., 2020). The reason for involving social sciences might be that the implementation of digital solutions in urban areas also implies attention to the social context in which it is implemented. Contextual aspects are also likely to influence how solutions are understood, used, and combined with other solutions. Finding out how urban areas can use technology and digital solutions for better[1] societal development therefore demands input from several disciplines. However, social sciences draw on different disciplines where the underlying assumptions about reality might differ. Dominantly, social science approaches to the smart city focus on substantive aspects, such as identifying best practices, barriers and enablers, or what the results of specific initiatives can be. *Smart city* as developing processes is often explored as linear processes, more or less de-coupled from the social situations through which they develop.

The aim of this chapter is to challenge the linear and substantive understandings of smart city initiatives and to present a processual and relational understanding of the smart city as emergent and evolving through social situations. A perspective embedded in American Pragmatism can provide a framework for understanding smart city 'solutions' not as fixed solutions isolated from its context but as elements that develop their meaning through the situations in which they are used. However, as this might seem overly abstract, it starts with an example, describing one of the basic elements – inclusion and co-creation – of the Norwegian Smart

DOI: 10.4324/9781003498650-8

City roadmap, to illustrate what linear and substantive understanding mean, before going on to presenting a processual and relational ontology.

Emphasis on inclusion and co-creation – an example of a linear and substantive understanding of the smart city concept

In the national roadmap for smart cities in Norway, developed by DOGA (Design and Architecture Norway; DOGA et al., 2019), emphasis on inclusion and co-creation is listed as one of eight main principles for realizing the smart city. The purpose of inclusion and co-creation is to increase citizen involvement and commitment to initiatives from the municipal and regional authorities (DOGA et al., 2019). However, there may be numerous challenges for doing so, such as lack of resources, competence, communication regarding processes, platforms and technology, or arenas and forums for said involvement and co-creation; internal resistance towards involving citizens; and absence of criteria for measuring the effect of co-creation and inclusion (DOGA et al., 2019). The questions that are understood as being key in solving such problematics are what platforms and digital tools can be used; where, when, and how to implement a system for co-creation so that it becomes an integrated part of public governance; who can facilitate these processes and with what tools; how to create better tailored solutions through co-creation and how to measure their effect; and how to spread knowledge, experiences, and solutions to other regions and municipalities (DOGA et al., 2019). However, what is drawn attention to in this chapter is that these best practices, tools, barriers, and enablers are not fixed. They all influence the situation in which they are used, as well as being influenced by the situation. Thus, if we want to achieve inclusion in an urban area by using a specific method, attention needs to be paid to more than the mere implementation of the method; that is, to what the method does to the ones being included and what those included do to the method as a co-constituting process.

The next part presents the core elements of a classical pragmatist understanding of reality, before going further into how this philosophy can be used to explore alternative understandings of the smart city concept and practice.

A radically processual understanding of reality

Realizing the smart and sustainable cities of the future will inevitably involve change, and change is fundamentally processual. However, process models and theories can be processual to different degrees, ranging from linear process models to what has been labelled as conjunctive process theories (Cloutier & Langley, 2020). An ontology and epistemology based in American Pragmatism can be categorized as conjunctive process thinking. The main difference between the conjunctive process theories and more linear process theories is that conjunctive process theories see *the process* as the unit of study, rather than seeing subjects – be they persons, material, or immaterial objects – as the unit of study. A linear process understanding typically addresses what kind of input is needed to produce a certain output. For example, what kind of tools are efficient for involving affected citizens in co-creating processes that also are seen as democratic? In linear process thinking, the input, throughput, and output factors are all understood as relatively static. Implicitly, linear process thinking does not take into consideration that both the understanding of the situation, tools, and participants will also develop through the development of the situation. In other words, it does not consider that the understanding of the process develops *through* the process.

Seeing the *smart city* in a more radical processual way rather than a substantive way means shifting attention from 'what' to 'how' (Rylander Eklund & Simpson, 2020). As an illustrative example, we can say that instead of searching for *what* the optimal bicycle-sharing solution in a city can be (see, for example, Müller-Eie, chapter 7 for a substantive understanding of bicycle sharing), we direct our attention to *how* the bicycle-sharing solution is interpreted and how it *impacts and is impacted by* users, stakeholders, and other aspects of city life. Further, how people use bicycles and how they respond to solutions also influences how other solutions develop and what possibilities can be imagined in the future. In a pragmatist ontology, the bike solution is seen as 'a better solution' for the situation here and now, rather than 'the best solution'. Hence, pragmatism is in its core experimental, where it is through practical action one explores ideas about what could possibly work, to learn more about both the situation and possible solutions. Pragmatism has a strong ethical dimension, striving to turn a current situation into a better one, not searching for the final solution but as a continuous activity of improving (Simpson & den Hond, 2021). However, that does not imply striving for a pre-defined and static 'better' but rather the constant striving of trying to improve the current situation into a better future.

Central elements of American Pragmatism

American Pragmatism was developed in the United States between the mid-19th and mid-20th centuries as a form of protest '… that rejects high-minded metaphysics in favour of understanding the everyday practicalities of living in an uncertain and ever-changing world' (Simpson & den Hond, 2021, p. 128). The main contributors to the development of this philosophical development were John Dewey, George Herbert Mead, Charles Sanders Peirce, and William James. They had different approaches, and although they knew each other and some of them (Mead and Dewey) also came to work closely together, it was probably never their intention to create a new 'school of thought called Pragmatism' (Simpson, 2017, p. 55). Two other scholars who have been influential in the development of American Pragmatism are Jane Addams and Mary Parker Follett. They were both working with social challenges through practical experimenting; Jane Addams, considered the mother of Social Work in the United States (Schneiderhan, 2011), and Mary Parker Follett through challenging existing ways of understanding democracy (see Follett, 1919) and bringing relational dimensions into Management thinking (see Follett, 1924).

There are different underlying assumptions of reality that we take for granted; they are therefore expressed to a much lesser extent, let alone questioned. Within management studies and social sciences, reality is typically seen as dualistic, thus making it challenging to understand solutions and the social situation they develop through as co-constituting. In a dualistic understanding, we see the individual and the social as two separate dimensions. Likewise, we separate the performer from the action being performed, the process from the result, inside vs. outside, stability vs. change, theory vs practice. These dualistic understandings address situations through their entities, seeing these entities as stable and separated from other entities. In contrast, seeing these dualisms as dualities implies seeing these entities as aspects of the same phenomenon (Simpson & den Hond, 2021). This shift from a dualistic understanding of reality to a duality understanding builds on six central elements where pragmatist ontology differs from a more metaphysical understanding of reality – namely: transactionality, temporality, habit, inquiry, meaning-making, and power – and its related pragmatist epistemology.

Transactionality

Seeing elements of situations as dualities rather than dualisms influences how we understand action. Dewey and Bentley (1949) differentiated between three onto-logical understandings of action. The first, called *intra-action*, sees action as something within and by the individual, where the acting individual is separate from her environment. Explanations of actions based on an individual's char-acteristics, abilities, and motivations are embedded in this ontological under-standing. In contrast, understanding action as something that goes on between individuals and groups is called *inter-action*. An understanding of inter-action starts with independently defined actors that remain the same through the inter-actions. Explanations of what happens are searched for in the acts, not the actors. Hence, both the understandings of intra-action and inter-action see acts and actors as dualisms. In contrast, a transactional understanding of action sees actions and actors as dualities, where actors both form and are formed by the actions. In other words, trans-action implies that actions both create movements in the par-ticipants and in the situation. This understanding also resonates with what Mead (1934) called 'conversations of gestures', where a gesture evokes some form of response in the other, and where the meaning is found in holding the gesture together with the response it calls out in the other. Consequently, it is not possible to see actions in a static way. Thus, the meaning of the situation and who the participants become in the situation is dynamically developing.

Temporality

A trans-actional understanding of reality is therefore also different from the understandings of intra-action and inter-action in terms of temporality. In intra-action, action happens within the person while the outside situation is seen as stable. In inter-action, action happens between persons, while there is no inner movement. A trans-actional understanding of reality implies that action happens both within and between those involved in the situation, implying that everything is in movement.

An ontological understanding embedded in American Pragmatism sees reality as temporally evolving, where both the past and the future are under continuous construction. This understanding of temporality is embedded in the work of Mead (1932), understanding reality as only existing in the present but where the interpretation of the present is informed by interpretations of the past and expectations towards the future. Such an understanding of reality differs from a Newtonian understanding of temporality. A Newtonian understanding of temporality sees time as linear, where the past is known to us, while the future is open (Hydle, 2015). The present is seen as a point in time, separating the known past from the unknown future. The way Mead (1932, 1934) saw it, in any present moment the emerging events can make us reconsider our interpretation of the past, which in turn also can lead to reconsiderations of our expectations towards the future.

Habit, inquiry, and meaning-making

Over time, we develop patterns of interaction, leading to habits. Habits help guide our everyday actions so that we in many ways act without much extra consideration. According to Simpson and Lorino (2016), Dewey understood habits as '… acquired and continuously modified through experience, but they never fully determine the course of action'. However, sometimes our assumptions about what will happen are subverted. In such a situation, doubt emerges and forces us to reconsider our previous understandings of the situation and develop alternative imaginations towards the future. Dewey and Bentley (1949) called these situations of doubt and reconsideration *inquiries* and understood them as situations where alternative understandings are searched for to make sense of what is happening. Thus, radically new understandings can emerge, creating new imaginations about the future.

Meaning-making (Simpson, 2009) is thus seen as a dynamic and social process that happens in a specific situation at a specific time. However, a 'situation' is not confined to a specific moment: it can last over time, dynamically developing. Meaning-making is thus a co-constitutive process between the development of events and those involved in the events. This understanding of meaning-making as a co-constitutive process of becoming is based on Dewey and Bentley's (1949) and Mead's (1932, 1934) understanding of temporality and transactionality (whereas Mead used the term *inter-actionality*).

Power

This understanding of temporality and transactionality, habits, and inquiries also implies a relational and processual understanding of power. Follett (1919, 1924) understood power as a social process. She described three ways of understanding power, the first being one having power *over others* through position or access to resources. Through this form of power, the needs of the ones exercising the power will overrule the needs of those with less power. The next form of power is through compromise, where a shared solution is found by considering all parties. However, this is not a solution that addresses the needs of any of the parties fully.

Follett (1919, 1924) advocated for a third way of understanding power, as a social and dynamic phenomenon where conflicting views and positions are expressed and integrated (Morlacchi, 2021). Through integration the participants develop ideas and understandings that none of them previously could imagine. It is through the different understandings of the same situation that the participants are able to see the situation in a new light and thereby re-interpret their own understandings in light of new dimensions. That does not, however, mean that participants reach an identical understanding of the situation. They develop individual versions of their shared understandings by taking the attitude of others from their own position.

As such, Follett's (1919, 1924) conceptualization of power resonates with Mead's (1934) understanding of gestural conversations and Dewey and Bentley's (1949) understanding of inquiry. When Follett (1924) talked about power through integration, it is the meeting of conflicting views that forces the participants in the situation to reconsider their own understanding through being exposed to another understanding of the same situation. This leads to movement in what is possible to imagine. This resonates with inquiry in the sense that it is the emergence of something different that forces participants to reconsider previous understandings of the situation and thus imagine other futures.

American Pragmatism developed in a time of major societal changes. The experience of societal changes and disruption that Follett (1919) described has parallels to our time. How can pragmatist ideas, ontology, and epistemology contribute to exploring contemporary questions connected to societal development and, more specifically, to the exploration of smart cities? An elaboration will follow on how the different aspects of *smart city* concepts can be understood from a pragmatist point of departure. First, what a pragmatist ontology can mean in terms of a pragmatist epistemology must be addressed.

A pragmatist approach and agenda for studying central aspects of the smart city

From a pragmatist approach, studying social situations is about the abductive process of inquiry. It is about hypothesizing what possibly could be, to test out whether current assumptions make sense and, further, re-interpreting both assumptions and hypotheses that conflict with current experiences so that assumptions align with current experiences. To draw attention to processual and relational aspects of smart city elements means shifting focus away from whether the specific solution or the result of a specific process works out well or not, to developing temporal hypotheses about how the solutions work on the situation and how the situation influences the solutions. Based in pragmatist ontology, the researcher is part of the social situation that is studied. It is through being part of the situation that the researcher can, to some extent, develop the attitude of the participants in the situation and see it from their angle. Through situations of doubt, both researchers and the involved participants are forced to reconsider current expectations and explore alternative imaginations. As such, the problem definition, the definition of the situation, and who the participants become in relation to one another are under continuous development. In an epistemology embedded in American Pragmatism, the questions that often are posed in smart city research become meaningless. There is a shift from addressing, for example, the implementation of a specific solution as a specific phase of the realization processes to seeing the development of a solution as interwoven with the implementation and, furthermore, as interwoven with the development of those affected by the solutions. This implies a shift in attention from what something is, to focusing on development in meaning.

The smart city as a concept has different dimensions, ranging from the implementation of digital structures, data security, and new digital business models to questions about how big data and the digitalization of society influence democracy, power, culture, and livelihoods. The smart city has also increasingly become a concept for legitimizing and labelling a city or urban area as a community at the forefront of urban development. Smart city concepts and roadmaps can give cities and city governments 'recipes' for how to realize the future society, by typically searching for best practices. From a pragmatist point of departure, it is interesting to study how the *smart city* concept influences how municipalities label their work with urban development, how it affects their collaboration with others and their organization of the work, and how their activities also influence their understanding of what smart cities can be. In the following paragraphs, central aspects of smart city research are commented on from a pragmatist approach.

Technology

Digital solutions are central tools for the smart city. Digital technology can be used for monitoring activity; allocating resources; providing information to public authorities, service providers, and citizens; and making services more convenient. Available technology can also push the utilizing of possibilities of the urban society. One example of this is digital twins, increasingly seen as a tool for both planning, monitoring, and providing services in the smart city. Embedded in a pragmatist understanding, attention is shifted from what tools *are* to what happens in transaction with the tools. Research on how digital twins can be used to make the city more accessible for tourists, for example, shifts from focusing on how the digital twin can provide more services to how the use of digital twins changes the way we interact, how tourists understand and use the city, and who those involved become in relation to one another and their surroundings. Hence, taking a pragmatist approach to technology means seeing technology as something that produces its meaning through transactions with users and stakeholders, and this meaning changes over time. It can be negotiated and challenged and can also change in a moment through an unexpected event. Technology also contributes to defining who those affected by experiencing it become, in relation to technology and to others related to the technology.

Best practice

Municipalities aiming to realize the smart city often develop best practices that can be implemented more widely (see Pozdniakova, 2018, as one example). Contextual differences might demand some adjustments, but the idea is that what has produced a positive result in a similar situation before can be repeated in a new situation. However, there are several challenges with this understanding. There is often no clear goal from the outset, making it challenging to 'measure' best practice (see Müller-Eie, chapter 7 for more elaboration on measuring of solutions). Seeing this from a pragmatist approach does not preclude the repeatability of good practice. On the contrary: habits are often informal forms of best practices. Nevertheless, habits are not fixed. They change according to changing circumstances. Likewise, habits also influence the circumstances. Hence, best practices should be understood as dynamic elements that influence and are influenced by the situations they are part of. Best practice as a concept also often becomes what Mead (1934) called 'significant symbols' (p. 47). Significant symbols are shared understandings in a group or society of certain phenomena, situations, or things. When leaders in organizations talk about developing best practices, leaders, employees, and others involved in the work have ideas about what that implies. They often have somewhat different expectations to what it implies for them, but the overall meaning is more or less shared. These shared understandings help the participants coordinate their work and have relevant expectations of what it can imply for their own situation. Becoming a professional leader or employee in the public sector implies understanding the meaning of best practices in terms of creating shared goals but also in terms of defining clear outputs and results of invested time and money. Studying best practice in smart cities from a pragmatist approach can mean directing attention to the developed understandings of what best practice is about and how it works.

Democracy, governance, and inclusion

As the *smart city* concept has developed from focusing more narrowly on how to use digital technology for making urban areas 'smarter', greater attention is now given to how smart cities become socially sustainable, fostering democracy and inclusion. Inclusion through co-creation and public innovation is expected to both foster democracy and empower marginalized groups (Ansell, 2011; Torfing, 2016; Vanleene et al., 2018; Voorberg et al., 2015). Nevertheless, this has proved to be difficult to do in practice. A main reason for this is possibly the understanding and handling of power. Follett (1919) presented three forms of power: domination, compromise, or integration. What would it mean for participants in public innovation to go from a more hierarchical understanding of power to power as integration? What consequences would it have for our understanding of responsibility and accountability? Could a pragmatist understanding of accountability as a situated and relational phenomenon (Kerveillant & Lorino, 2021) contribute to alternative understandings of what is possible to do? A pragmatist understanding can open up possibilities for those affected, simultaneously being those who affect. One of the goals of co-creation is to empower those affected by the situation. However, that would imply that both people and the situation can change through the process. A pragmatist approach opens for a logic where that can happen.

Implementation and assessment

Research on the smart city has also been especially concerned with how technological smart city solutions are implemented in the best or most effective way (as two of many examples, see Angelakoglou et al., 2019; Orlowski & Romanowska, 2019). This also implies an assessment of the implementation. The assessment of success is typically measured through comparing the outcomes and results of a given implementation to the initial goals. However, acknowledging the fact that circumstances change, we learn along the way, and we reconsider situations as they develop, would it not be better to understand the realization of prospective solutions as on-going processes of exploring what meaning the prospective solutions can have in the current situations and adjusting the aims and intentions in relation to that? A pragmatist understanding of realizing solutions will not understand intentions, implementation, and assessment as phases of sequential elements. They are rather aspects that co-construct one another continuously. In other words, while realizing solutions with the intention of achieving a certain outcome, participants continuously reconsider the situation by both judging previous expectations compared to the current situation and reconsidering its aims and prospective actions (see Lindland and Matre, chapter 15 for how the meaning of a strategy changes through implementation). Deviating from initial intentions is thus not an exemption but rather the process of reflecting upon what happens and adjusting to the emerging situation.

Conclusion

The *smart city* concept can be addressed by emphasizing numerous themes, such as technology, planning, economic, governance, and social inclusion. The aim of this chapter has been to present a processual understanding of reality, embedded in American Pragmatism, and to show how this can contribute to an alternative research agenda for exploring the *smart city* concept in a manner closer to practice.

Pragmatist perspectives on the smart city can contribute to a more nuanced and deeper understanding of the needs, motivations, and social dynamics of urban development. This can contribute to developing our ability as researchers and practitioners to take the attitude of others in imagining future possibilities.

Taking a pragmatist approach implies a shift in ontology from seeing events as detached, individual, linear, and static to co-constituting meaning-making processes of becoming. This does not mean that events and facts are disregarded but rather that our interest shifts from what has happened to what the events do to the situation and those involved in the situation. Making this shift can be challenging, as we are predominantly trained in a substantive way of understanding reality. It means that we, for example, will find the idea of 'best practice' less productive or meaningful, as any 'practice' will be realized in a social situation that will influence and be influenced by the situation, making it different from how it was foreseen at the time it was introduced.

Implications

Conducting research embedded in a pragmatist approach implies seeing oneself as a researcher as part of the social situation that is researched. There is no outside position the researcher can take. It is through developing the attitude of those experiencing the situation that the researcher can possibly imagine how those affected experience the situation.

Smart city concepts can both legitimize and make space for alternative ways of working with urban development across organizational boundaries. However, in situations where the aim is co-creation and inclusion for making more democratic work processes, it is also important to address the issues that are of major concern for the people involved. People will engage themselves in the issues they find important to them. Nevertheless, these issues are not necessarily the same issues that they are invited to collaborate on. Co-creation and co-production of future urban solutions could thus also come from stakeholders and citizens, not just smart city project owners.

Some suggestions for further research on smart cities from a pragmatist perspective include the following:

- How is democratic inclusion operationalized in smart city co-creation and co-production with citizens and other stakeholders?
- How does the implementation of smart city concepts change the way municipalities work with urban development?
- How can a pragmatist approach to accountability redirect attention towards what happens in smart city activities?

Note

1 What is meant by 'better' can be many-facetted. It can be related to better economic solutions, better accessibility, better coordination, and/or better social inclusion.

References

Angelakoglou, K., Nikolopoulos, N., Giourka, P., Svensson, I.L., Tsarchopoulos, P., Tryferidis, A., & Tzovaras, D. (2019). A methodological framework for the selection of key performance indicators to assess smart city solutions. *Smart Cities*, 2(2), 269–306.

Ansell, C. (2011). *Pragmatist democracy: Evolutionary learning as public philosophy*. Oxford University Press.

Cloutier, C., & Langley, A. (2020). What makes a process theoretical contribution? *Organization Theory*, 1, 1–32.

Dewey, J., & Bentley, A. (1949). *Knowing and the known. The later works 1949–1952*. Southern Illinois University Press.

DOGA, Norwegian Smart City Network & Nordic Edge. (2019). *Roadmap for smart and sustainable cities and communities in Norway. A guide for local and regional authorities developed by Design and Architecture Norway (DOGA), the Norwegian Smart City Network and Nordic Edge.*

Echebarria, C., Barrutia, J.M., & Aguado-Moralejo, I. (2021). The smart city journey: A systematic review and future research agenda. *Innovation: The European Journal of Social Science Research*, 34(2), 159–201.

Follett, M.P. (1919). Community is a process. *The Philosophical Review*, 28(6), 576–588.

Follett, M.P. (1924). *Creative experience* (Vol. 7). Longmans, Green and Company.

Hydle, K.M. (2015). Temporal and spatial dimensions of strategizing. *Organization Studies*, 36(5), 643–663.

Kerveillant, M., & Lorino, P. (2021). Dialogical and situated accountability to the public. The reporting of nuclear incidents. *Accounting, Auditing & Accountability Journal*, 34(1), 111–136.

Mead, G.H. (1932). *The philosophy of the present*. Philpapers.org.

Mead, G.H. (1934). *Mind, self and society* (Vol. 111). University of Chicago Press.

Morlacchi, P. (2021). The performative power of frictions and new possibilities: Studying power, performativity and process with Follett's pragmatism. *Organization Studies*, 42 (12), 1863–1883.

Orlowski, A., & Romanowska, P. (2019). Smart cities concept: Smart mobility indicator. *Cybernetics and Systems*, 50(2), 118–131.

Pozdniakova, A.M. (2018). Smart city strategies 'London-Stockholm-Vienna-Kyiv': In search of common ground and best practices. *Acta Innovations*, 27, 31–45.

Rylander Eklund, A., & Simpson, B. (2020). The duality of design(ing) successful projects. *Project Management Journal*, 51(1), 11–23.

Schneiderhan, E. (2011). Pragmatism and empirical sociology: The case of Jane Addams and Hull-House, 1889–1895. *Theory and Society*, 40(6), 589–617.

Simpson, B. (2009). Pragmatism, Mead and the practice turn. *Organization Studies*, 30 (12), 1329–1347.

Simpson, B. (2017). Pragmatism: A philosophy of practice. In C. Cassell, A.L. Cunliffe, & G. Grandy (Eds.), *The SAGE handbook of qualitative business and management research methods: History and traditions* (pp. 54–68). Sage.

Simpson, B., & den Hond, F. (2021). The contemporary resonances of classical pragmatism for studying organization and organizing. *Organization Studies*, 43(1), 127–146.

Simpson, B., & Lorino, P. (2016). *Re-viewing routines through a pragmatist lens*. Strathprints.strath.ac.uk

Torfing, J. (2016). *Collaborative innovation in the public sector*. DJØF Publishing.

Vanleene, D., Voets, J., & Verschuere, B. (2018). The co-production of a community: Engaging citizens in derelict neighbourhoods. *VOLUNTAS: International Journal of Voluntary and Nonprofit Organizations*, 29(1), 201–221.

Voorberg, W.H., Bekkers, V.J., & Tummers, L.G. (2015). A systematic review of co-creation and co-production: Embarking on the social innovation journey. *Public Management Review*, 17(9), 1333–1357.

Zheng, C., Yuan, J., Zhu, L., Zhang, Y., & Shao, Q. (2020). From digital to sustainable: A scientometric review of smart city literature between 1990 and 2019. *Journal of Cleaner Production*, 258, 120689.

6 Towards the streetsmart city

A research and planning agenda for inclusive cities

Jens Kaae Fisker, Anders Riel Müller and Helene Eiliott

From exclusive to inclusive cities

Despite a well-established international consensus in the United Nations' New Urban Agenda that our cities ought to be for everyone (United Nations, 2017), too often they are planned, designed, built, managed, and regulated primarily for the benefit of small, privileged segments of the population. In recent decades, much of this can be attributed to the financialization of urban space and the powerful positions enjoyed by private developers looking to make a profit (Fields, 2017; Mosciaro, 2021). As a result, gentrification has intensified, with large swathes of urban space becoming inhabitable only for those affluent enough to live there. The broader problem, however, extends to other axes of social differentiation and is not limited to the specific issue of gentrification. Urban scholars have shown how the cities we have inherited were largely made by and for men (e.g., Kern, 2020) within a largely heterosexist context (e.g., Kitchin & Lysaght, 2003), how gentrification tends to be ethnically and racially skewed (e.g., Goetz, 2011; Kirkland, 2008) while also disproportionally displacing working-class subjects (e.g., Paton, 2014), and how the ableist city tends to ignore or neglect the needs of inhabitants with disabilities (e.g., Parent, 2018). Together, these (and other) aspects of urban development contribute to the (re)production of *exclusive cities* and thereby stand in the way of realizing the goal of *inclusive cities* as set out in both the UN's *New Urban Agenda* (2017) and the EU's New European Bauhaus programme.[1]

The issue at hand is by no means new. In the early years of critical urban theory, David Harvey (1973, p. 27) pointed out that 'once a particular spatial form is created it tends to institutionalize and, in some respects, to determine the future development of social process'. In other words, when urban planners and private developers co-create urban spaces, they also funnel both development trajectories and the everyday uses of the city in particular directions. When urban spaces are designed for certain kinds of people, functions, activities, and so forth, it will have implications for how those spaces are used and by whom. This does not mean, however, that the uses of urban space are *determined* by urban planning and regulation; *conditioned* yes, but not determined. Urban inhabitants are also guided by various socio-cultural norms and furthermore retain the possibility of using the city

DOI: 10.4324/9781003498650-9

in ways that were not intended by design. Likewise, they may challenge social norms by doing unexpected things in unexpected places. Still, the power of urban planning and design to influence how urban lives are lived should not be underestimated, and neither should their potential for enacting change:

> We have the opportunity to create space, to harness creatively the forces making for urban differentiation. But in order to seize these opportunities we have to confront the forces that create cities as alien environments, that push urbanization in directions alien to our individual and collective purpose.
>
> (Harvey, 1973, 313f)

The forces that Harvey had in mind were primarily those of a capitalism, whose impacts on urban life were to make cities increasingly fit for generating profits but decreasingly fit for human habitation – a point forcefully reiterated 4 decades later in the sloganized title *Cities for People, Not for Profit* (Brenner et al., 2012). Moreover, even when attempts are made to create cities for people, the result may be cities for *some* people and not for others. In some cases, exclusionary targeting has even been conducted openly as exemplified by the global quest to attract the so-called creative class (Peck, 2005).

The arrival on the scene of the smart city, and in particular the increased use of big data, algorithms, and automation, shows signs of compounding the existing feedback mechanisms through which urban exclusion and inequality are reproduced and exacerbated (Barns, 2021; Vanolo, 2014; Willis, 2019). As Sarah Barns (2021, p. 3208) asked: 'If big data visualises, replicates and reproduces that which is routine about existing cities, and uses these existing patterns to create agents that act autonomously on behalf of humans, what prospects are there for change?' Adding another layer of concern, the smart city technologies being rolled out were developed for the presumed needs and preferences of a very limited segment of people, meaning that exclusion is already built in to big data (Cardullo & Kitchin, 2019; Shelton & Lodato, 2019). Research on smart cities has also shown a tendency towards centralized, top-down implementation where real participation and co-creation are precluded (Rogan, 2019; Späth & Knieling, 2018), while developments are largely driven by corporate rather than citizen interests (Cowley & Caprotti, 2019; Marvin & Luque-Ayala, 2017; Rose, 2020). Levels of surveillance increase (Mosco, 2019; Offenhuber, 2019; Wood & Mackinnon, 2019), especially for the most vulnerable populations (Coleman, 2020). In addition, scholars in the Global South have shown how smart city projects colonize the future of cities (Datta, 2019). In short, smart cities have reproduced existing inequalities and created new ones (Vanolo, 2014; Willis, 2019).

Simultaneously, urban planning has been shown to be deeply involved in enacting the financialization of urban space (Savini & Aalbers, 2016; Waldron, 2019), and to reproduce heteronormative assumptions in its practices and hence in its outputs (Doan, 2011). Taken together, what may emerge is a situation where urban planning processes and smart city technologies actively reinforce one another, creating a feedback loop where collected data are fed into the planning process, generating new exclusionary urban spaces, which further exacerbate the patterns picked up by smart city technologies in the first place (Willis, 2019). This makes it even more important for planning practices to be designed in ways that allow them to break such emerging loops of urban inequality and exclusion rather than reproduce them. While urban planners and designers cannot be held solely responsible for the (re)production of exclusive cities, they have certainly been complicit in the developments mentioned above. The flipside of the fact that planning practice has been part of the problem is that it can also be part of the solution.

In this chapter we address the following challenges: how can we develop and deliver planning and research practices that contribute to the production of inclusive rather than exclusive cities? And what roles should urban planners, artists, researchers, and technology developers play in the production of inclusive cities? We discuss these questions by engaging with relevant strands of the academic literature and outline how this has informed the participatory processes in the New European Bauhaus Stavanger (NEB-STAR) project, where some aspects of the challenge are being explored through experimental practice. The project follows up on the City of Stavanger's participation in the EU Smart City Lighthouse project Triangulum where the focus was on technological innovation and demonstration for smart sustainable cities (Haarstad & Wathne, 2019). NEB-STAR is one of six lighthouse demonstrator projects in Europe funded by the European Union. Stavanger is the lighthouse city in the NEB-STAR project, and two neighbourhoods of Stavanger have been selected for demonstration projects that experiment with participatory planning processes aimed at fostering more sustainable, beautiful, and inclusive urban development (Stavanger Municipality, 2023).

The right to the smart city

In response to the situation outlined above, calls for a 'right to the smart city' have been made, aimed at developing principles and practices capable of promoting and safeguarding social justice, equality, and inclusion (Cardullo et al., 2019; Foth et al., 2015; Willis, 2019). The current chapter responds to these calls by proposing methods for enacting such a right through research and planning practice. First, however, we need to clarify what a *right to the smart city* entails in principle as well as in practice. Henri Lefebvre developed the concept in the context of 1960s Paris, where his writings on the subject coincided with and helped shape the events of May 1968 (Schmid, 2012). It is important to understand that this is not an individual but a collective or common right which 'gathers the interests [...] of the whole society and firstly of all those who *inhabit*' (Lefebvre, 1996, p. 158). It is also not merely a right to be present in urban space but 'a transformed and renewed *right to urban life*' (Lefebvre, 1996, p. 158), which includes alongside habitation also the right to participate in managing and creating the city (Fernandes, 2017; Purcell, 2008). In this sense, 'to inhabit the city' means more than simply residing in it:

The right to the city demands that we see the city first and foremost as *inhabited*. The city is a collective creative work by and for inhabitants. It depends on them, for they are its creative producers. But they also depend on the city: it is their habitat, the space of their everyday survival. Nurturing the relationship between city and inhabitant, therefore, [...] becomes the principal imperative of urban politics.

(Purcell, 2008, p. 105)

Following calls for the right to the city thus means prioritizing the city-as-inhabited over the city-as-property (Purcell, 2008). It also places inclusion and participation in urban decision making front and centre, not only as a means to an end but also as an end in its own right. Inclusion and participation in the context of the right to the city is a normative rather than an instrumental choice. In practice, however, making that choice in a manner that takes effect is fraught with difficulty. For instance, in Norway the legally binding nature of plans tends to clash with the societal mandate of municipalities.

The smart city has thrown up both challenges and new possibilities for pursuing the right to the city, of which three relevant ones are outlined here. First, the addition of digital urban spaces compels us to rethink inhabitation as extending also to the digital realm, something that has gained unprecedented visibility during the pandemic when many people were cut off from using physical public space for extended periods (Mullick & Patnaik, 2022). This evokes questions about how the physical and digital spaces of the city are inhabited, remembering that to inhabit is both to use and to create. The interface between the physical and digital realms becomes a key point of attention for both research and planning, with technologies such as augmented reality blurring the boundaries between the two. Second, urban data become a central element of concern: Who owns it? Who controls it? Who can access and use it? What is datafied and what is not? What are the legitimate uses of which data? Ultimately, these are 'normative questions about the goals of data-driven urbanism and whose interests they should serve' (Kitchin et al., 2018, p. 11). Here, we need to return to the priority given to inhabitation over property: this extends to urban data. Within a right to the smart city framework, the legitimacy of collecting and using such datasets resides in whether or not it is in the interests of inhabitants. Third, the utilization of big data as an input for regulation and planning, as mentioned earlier, risks creating and closing a feedback loop that would generate a vicious circle of exclusion and inequality. Safeguarding against this risk has to be a key concern for advocates of a right to the smart city.

Participatory planning revisited

Creating inclusive cities by way of realizing the right to the smart city is not a project that has to start from scratch. Participatory planning was introduced precisely to create more inclusive cities, where citizens were invited to co-create the outcomes of urban planning. The problem of reaching marginalized and vulnerable populations, however, has yet to be adequately solved. Public hearings and other conventional tools of participatory planning are often capable of reaching only a select population of resourceful, well-educated citizens who tend to resemble the socioeconomic profiles of planners themselves (Albrechts et al., 2019; Umemoto, 2001). As a result, inclusion is achieved only on paper through tokenistic practices (Monno & Khakee, 2012), leading to limited success in influencing urban development (Agger & Larsen, 2009). For some urban scholars, this represents a crisis of participatory planning (e.g., Purcell, 2016; but see also Legacy, 2017).

Several policy initiatives have acknowledged participatory shortcomings in smart city projects. The national *Roadmap for Smart and Sustainable Cities and Communities in Norway* places participation and inclusion at the centre (Design and Architecture Norway, 2019). Likewise, participation and co-creation are highlighted in the European Mission on *Climate-Neutral and Smart Cities* (European Commission, 2020), which continue the (at least symbolic) shift from a techno-centric to a citizen-centric approach to smart cities (cf. Calzada, 2018). The New European Bauhaus programme has further emphasized inclusion by including it as one of three guiding values along with aesthetics and sustainability (European Commission, 2022). Finally, the United Nation's *New Urban Agenda* (2017) makes a commitment to developing smart cities, including the use of platforms and citizen-centred digital governance tools for broadening participation and ensuring responsible governance practice. It also positions participation and inclusion as a key priority, especially where the most vulnerable urban dwellers are concerned (United Nations, 2017). There is, however, little to no indication in these policies about how to overcome the limitations of prevailing participatory methods. Developing new methods is therefore crucial in the context of the emerging crisis in participatory planning, where participation is often reduced to a tokenistic practice of legitimizing decisions, rather than an open-ended process where the knowledge and ideas of citizens are taken seriously (Legacy, 2017; McAuliffe & Rogers, 2018).

Based on Arnstein's (1969) classic ladder of participation, Cardullo and Kitchin (2019) have helpfully proposed a 'scaffold of smart citizen participation' as a means for analysing the supposedly citizen-centric turn in smart city projects. The scaffold (see Table 6.1) is useful not only for analytical purposes but also as a tool for designing participatory planning practices for various purposes, including not least the sort called for in this chapter. Ultimately, the right to the smart city entails a pursuit of participatory methods that fall within the 'citizen power' level of participation, covering 'citizen control', 'delegated power', and 'partnerships'. But this does not mean that activities belonging elsewhere on the scaffold should simply be excluded from consideration. Their usefulness and legitimacy depend on the wider constellation of methods within which they are set. If 'consultation' is the only form of participation being employed, then its categorization as 'tokenism' seems correct. But if 'consultation' activities are instead used, for instance, as a preparation for setting up 'partnerships' where inhabitants and planners co-create, then the conclusion has to be different. So, the practical usefulness of the scaffold is to ensure that methods are being used as part of a wider setup, where the overall outcome can be categorized as 'citizen power'.

Table 6.1 The scaffold of smart citizen participation

Form and level of participation		Role	Citizen involvement	Political framing	Modality
Citizen power	Citizen control	Leader, member	Ideas, vision, leadership, ownership, create	Rights, social/political citizenship, commons	Inclusive, bottom-up, collective, autonomy, experimental
	Delegated power	Decision-maker, maker			
	Partnership	Co-creator	Negotiate, produce	Participation, co-creation	
Tokenism	Placation	Proposer	Suggest		Top-down, civic paternalism, stewardship, bound to succeed
	Consultation	Participant, tester, player	Feedback	Civic engagement	
	Information	Recipient	Browse, consume, act		
Consumerism	Choice	Resident, consumer		Capitalism, market	
Non-participation	Therapy	Patient, learner, user, product, data point	Steered, nudged, controlled	Stewardship, technocracy, paternalism	
	Manipulation				

Source: Adapted from Cardullo and Kitchin (2019, p. 5).

From smart to streetsmart

The scaffold of participation (Cardullo & Kitchin, 2019) provides an important tool for analysis and evaluation, but it cannot provide prescriptions for how to create more inclusive cities. What it makes clear is that a truly inclusive smart city requires more than the collection and valorization of citizen knowledge in all its diversity. It requires a sharing of the decision-making powers usually vested in politicians, planners, and property owners. If we take the right to the smart city seriously, then the whole logic of participation needs to be inverted. In the current situation, citizens participate in a planning process 'owned' by planners, politicians, and property owners. Ideally, this needs to be inverted to create a situation where these powerful groups are instead invited to participate in a planning process controlled by urban inhabitants assisted by relevant experts such as planners, artists, academics, and architects. In a general sense, such a situation is not attainable without major legislative reforms, but we argue that even within current planning regimes movement in this direction is possible.

By introducing the term *streetsmart city*, we wish to signal that our role as academic researchers is to enable the knowledge and visions of inhabitants – and especially those of vulnerable populations – rooted in lived experience, to gain political potency in the urban planning process; in short, to develop cities that start at the street level and reflect the everyday realities and desires of vulnerable populations and hence enable more inclusive urban futures (cf. Datta, 2019; Fisker et al., 2019). This requires us to bring the lived experiences and street-level knowledges (including affect and emotion) of these populations into the formal planning process (Escalante & Valdivia, 2015) and to ensure that planning outcomes reflect these experiences and knowledges. Previous research has tended to study cases that are either firmly within the boundaries of the planning system (e.g., Monno & Khakee, 2012) or decidedly beyond it (e.g., Purcell, 2016). While these produce valuable knowledge, we also need studies that seek to explore the *limits* of the planning system by pushing at the boundaries of what is possible in the absence of major legislative reform. Research and planning need to grapple with three interrelated questions to do so:

1   *How can planners, researchers, artists, and vulnerable populations actively work together to reverse power asymmetries that exclude and marginalise minority perspectives on urban space?* Vulnerable populations possess lived knowledge of urban space which is usually not valorized and taken seriously in the same way as 'expert' knowledge. This requires us to challenge and unsettle the taken-for-granted positionalities and roles of different stakeholders while working proactively with the asymmetrical power relations that exist among them (Gibson-Graham, 2006). The task is to transform power differentials through collaborative practices to give vulnerable populations access to the legitimating expertise of planners, artists, researchers, and smart city technologies to co-produce political potency at various strategic points in the planning process. As such, we move away from an 'extractive' model of research towards co-production for social change.

2   *How can researchers, artists, and urban planners rethink their own role and positions of power in urban planning?* Urban planners, academic researchers, and artists are broadly perceived as experts in their respective fields – planners and politicians in political–administrative decision-making processes, researchers on the production of scientific knowledge, and artists as experts in aesthetic representations. The political legitimacy of such expertise positions them as privileged stakeholders in the policy and planning system. The question requires us to acknowledge our role as privileged stakeholders, to determine how we use our roles as experts and critically engage with how our expertise is used politically.

3   *How can digital technologies such as augmented and virtual reality, Digital Twin,[2] and participatory decision-making tools be recalibrated to enable and promote the inclusion of vulnerable populations?* This entails an exploration of the extent to which smart city technologies currently in use are inherently biased towards maintaining or reinforcing existing power asymmetries and to what extent they can be 'recalibrated' to reverse asymmetries (Cardullo et al., 2019). As Ferguson (2010, p. 182f) argued, 'It is not the machines or the mechanisms that decided what they will be used to do', but still, the design of technologies does usually impose some constraints on use. A significant challenge in this context is the digital divide, which tends to preclude some vulnerable groups from participation, either because they lack technology access or because of variations in digital literacy (e.g., Colding et al., 2024; Jang &

Gim, 2022). Previous research suggests that this can be partially overcome through technology design (Giannoumis & Joneja, 2022; Kirklies et al., 2024), but more work is needed on different ways of using available technology in more accessible ways.

By focusing not only on urban planners, we suggest that inspiration for relevant participatory methods can be drawn also from research, architecture, and artistic practice. Vasudevan (2020, p. 70) argued that using art can 'help relieve power imbalances and communication barriers prevalent in traditional communicative and participatory processes' because the 'non-verbal nature of art production actually facilitates greater communication between the researcher and participant, especially if the quality of the art is de-emphasized and reflection, expression, and communication is emphasized'. Furthermore, the use of art for participation can be conducive for memory and emotion and thus help 'to undo the separation between cognitive objectivity and emotional sensitivity in plan-making' (Vasudevan, 2020, p. 70). Urban planning is as much about emotional attachments to place as it is about rational arguments. Safeguarding a right to the smart city therefore requires methods that valorize and capture emotions and affect without subordinating them to conventional planning rationalities.

In the action research tradition, the (Utopian) Future Workshop has a long history of use, including in a planning context (e.g., Kesselring & Freudendal-Pedersen, 2021; Khakee, 2010; Pluchinotta et al., 2019). It structures the participatory process into three phases: critique, utopian, and realization phases (Jungk & Mullert, 1998). In the version implemented in the NEB-STAR project, the *critique* phase is used to identify shortcomings of existing urban spaces as seen from the perspective of various vulnerable, marginalized, and excluded groups (Eiliott et al., 2023). The first part of the *utopian* phase is then used to envision the qualities of alternative urban spaces where those shortcomings have been remedied. In the second part of the utopian phase, these qualities provide the basis for planning inputs and artistic and digital productions during the *realization* phase. The phases are workshop-based with researchers and artists playing key roles as facilitators. Researchers attempt to bring forth knowledge and visions while artistic researchers and artists work with participating inhabitants to develop visual representations of both critique and utopian visions. This is important because aesthetic representations are politically powerful in planning processes as exemplified by the importance of architectural drawings. The realization phase aims at converting visions into realizable action plans and proposals. Timewise, the realization phase is the longest as it includes the lengthy process of providing input to urban development plans, political negotiation, and plan revisions. This is a reiterative process in which the suggested plans are revised by revisiting and reformulating the outputs of the critique and utopian phases. The role of researchers, artists, and planners is to lend their roles as experts to uphold political pressure and ensure that the knowledge, visions, and planning inputs put forth remain politically potent. This is a crucial phase where vulnerable populations, researchers, artists, and planners must work together, using their roles as 'experts' strategically to ensure that the visions developed do not become sidelined in the planning process. The realization phase is where the potentials and limitations of the collaboration are put to the test.

Conclusion

The key objective of this research and planning agenda is to maximize the political potency of marginalized visions in urban development. By *political potency* we mean to move beyond symbolic participation by leveraging the legitimating powers and expertise of researchers, artists, and planners strategically in urban planning. The agenda would have three ancillary objectives: (1) to develop participatory methods that are designed to bring the knowledge and experiences of vulnerable populations to the forefront of urban planning; (2) to critically explore and rework the role of artists, researchers, and urban planners in the urban planning process in ways that will lead to just and inclusive urban futures; and (3) to explore the potential and limitations of smart city technologies in envisioning and realising just and inclusive urban futures.

Such work can, for instance, draw on knowledge developed in anti-racist literature on 'the ally' (e.g., Case, 2012; Utt, 2021; Williams & Sharif, 2021). This entails an exploration of the extent to which power asymmetries can be addressed within existing formalized planning processes. The choice of such a narrow scope is predicated on the fact that changing formal planning procedures is difficult without legal reform. Such reform may be necessary, but we want to focus here on the possibilities that exist within the confines of the current system. This should make it more difficult to find excuses for not doing what is already possible (among planners and researchers alike). In short, we seek to gain knowledge about the possible 'room for improvement' within the structural constraints of current planning laws. We propose that this can be done by combining participatory methods from art (Eynaud et al., 2018; Iannelli & Musaro, 2017; Vasudevan, 2020; Westenberg & Rutten, 2017), planning (Escalante & Valdivia, 2015; Horelli, 2017; Hudson & Rönnblom, 2020), and research (Engeström & Pyörälä, 2021; Englund & Price, 2018; Gibson-Graham, 2006; Jungk & Mullert, 1998; Schwencke, 2017), as well as using new smart city technologies in support of marginalized visions. This leveraging of technologies and various forms of expertise can be directed at strategic points throughout the planning process.

In the NEB-STAR project, we seek to break new ground by putting the principles of just, inclusive, and participatory decision making at the centre of the project instead of technology – not just nominally but substantively. Whereas participatory research, participatory art, and participatory planning are well-established fields, we propose to mobilize all three strategically along with digital technologies. We want to develop new knowledge about how different participatory methods and smart city technologies in combination, rather than in isolation, can open pathways for more just and inclusive cities. Thereby, we also hope to pave new roads for exploring the potentials and challenges for just participation and inclusion in public decision-making processes in general. We thereby address the

European Union's call for public engagement and citizen science and the co-production of knowledge but also recognize the crisis of participation as a sore point. We want to explore the potentials of citizen science by redefining the roles and reversing the object of study: from the study of marginalized and vulnerable populations' participation to that of the role of science, art, technology, and planning processes as enablers for these populations to determine urban development – or not (cf. Gibson-Graham, 2006). Our position is that 'expert' knowledge on the qualities of urban development resides with the people who live in cities – smartness should be sought in the streets. Therefore, we wish to push boundaries for scientific knowledge by co-producing scientific and artistic outputs with participants. In this regard, we seek to break down barriers and asymmetrical power structures between different regimes of knowledge and expertise.

The NEB-STAR project opens opportunities for participatory experimentation in a city where participatory planning has been rather limited historically. The outcome of this experimentation remains to be seen as the project is only half-way at the time of writing. It is thus too early to reflect on outcomes, successes, and failures. These forms of experimentation spaces where new forms of participatory planning in conjunction with new digital technologies can occur are, however, essential if we are to overcome the crisis of participation in urban planning.

Notes

1 https://new-european-bauhaus.europa.eu/index_en
2 In a smart city context, digital twins are attempts to make a digital copy of a city or neighbourhood which can then be used for purposes of simulation, testing, monitoring, visualization, etc.

References

Agger, A., & Larsen, J.N. (2009). Exclusion in area-based urban policy programmes. *European Planning Studies*, 17(7), 1085–1099.

Albrechts, L., Barbanente, A., & Monno, V. (2019). From stage-managed planning towards a more imaginative and inclusive strategic spatial planning. *EPC: Politics & Space*, 37(8), 1489–1506.

Arnstein, S.R. (1969). A ladder of citizen participation. *Journal of the American Institute of Planners*, 35(4), 216–224. doi:10.1080701944366908977225

Barns, S. (2021). Out of the loop? On the radical and the routine in urban big data. *Urban Studies*, 58(15), 3203–3210.

Brenner, N., Marcuse, P., & Mayer, M. (Eds.). (2012). *Cities for people, not for profit: Critical urban theory and the right to the city*. Routledge.

Calzada, I. (2018). Deciphering smart city citizenship. *Revista Internacional de Estudos Vascos*, 63(1–2), 44–83.

Cardullo, P., Feliciantonio, C., & Kitchin, R. (Eds.). (2019). *The right to the smart city*. Emerald.

Cardullo, P., & Kitchin, R. (2019). Being a 'citizen' in the smart city: Up and down the scaffold of smart citizen participation in Dublin, Ireland. *GeoJournal*, 84(1), 1–13. doi:10.1007/s10708-018-9845-8

Case, K.A. (2012). Discovering the privilege of Whiteness: White women's reflections on anti-racist identity and ally behavior. *Journal of Social Issues*, 68(1), 78–96.

Colding, J., Nilsson, C., & Sjöberg, S. (2024). Smart cities for all? Bridging digital divides for socially sustainable and inclusive cities. *Smart Cities*, 7(3), 1044–1059.

Coleman, R. (2020). Surveillance as an urban way of life. In Z. Krajina & D. Stevenson (Eds.), *The Routledge companion to urban media and communication* (pp. 194–203). Routledge.

Cowley, R., & Caprotti, F. (2019). Smart city as anti-planning in the UK. *EPD: Society and Space*, 37(3), 428–448.

Datta, A. (2019). Postcolonial urban futures. *EPD: Society and Space*, 37(3), 393–410.

Design and Architecture Norway. (2019). *Roadmap for smart and sustainable cities and communities in Norway*. Norwegian Smart City Network.

Doan, P. (Ed.). (2011). *Queerying planning: Challenging heteronormatice assumptions and reframing planning practice*. Routledge.

Eiliott, H., Kverneland, K., Fisker, J.K., & Ahlers, D. (2023). *Co-creation process for NEB Stavanger (initial version)*. https://nebstar.eu/reports/d1-1-co-creation-process-for-neb-stavanger-initial-version/

Engeström, Y., & Pyörälä, E. (2021). Using activity theory to transform medical work and learning. *Medical Teacher*, 43(1), 7–13.

Englund, C., & Price, L. (2018). Facilitating agency. *International Journal for Academic Development*, 23(3), 192–205.

Escalante, S.O., & Valdivia, B.G. (2015). Planning from below. *Gender & Development*, 23 (1), 113–126.

European Commission. (2020). *Climate-neutral and smart cities: 100 Climate-neutral cities by 2030 – By and for the citizens*.

European Commission. (2022). *New European Bauhaus compass*.

Eynaud, P., Juan, M., & Mourey, D. (2018). Participatory art as a social practice of commoning to reinvent the right to the city. *Voluntas*, 29, 621–636.

Ferguson, J. (2010). The uses of neoliberalism. *Antipode*. 41, 166–184.

Fernandes, E. (2017). 'Right to the city' and the new urban order. In A. Brown (Ed.), *Rebel streets and the informal economy* (pp. 51–62). Routledge.

Fields, D. (2017). Urban struggles with financialization. *Geography Compass*, 11(11), e12334.

Fisker, J.K., Chiappini, L., Pugalis, L., & Bruzzese, A. (Eds.). (2019). *Enabling urban alternatives*. Palgrave.

Foth, M., Brynskov, M., & Ojala, T. (Eds.). (2015). *Citizen's right to the digital city*. Springer.

Giannoumis, G.A., & Joneja, N. (2022). Citizen participation and ICT for urban development in Oslo. In S. Hovik, G.A. Giannoumis, K. Reichborn-Kjennerud, J.M. Ruano, I. McShane, & S. Legard (Eds.), *Citizen participation in the information society: Comparing participatory channels in urban development* (pp. 97–116). Springer.

Gibson-Graham, J.K. (2006). *A post-capitalist politics*. University of Minnesota Press.

Goetz, E. (2011). Gentrification in Black and White: The racial impact of public housing demolition in American cities. *Urban Studies*, 48(8), 1581–1604.

Haarstad, H., & Wathne, M.W. (2019). Smart cities as strategic actors: Insights from EU lighthouse projects in Stavanger, Stockholm and Nottingham. In A. Karvonen, F. Cugurullo, & F. Caprotti (Eds.), *Inside smart cities: Place, politics and urban innovation* (pp. 102–116). Routledge.

Harvey, D. (1973). *Social justice and the city*. Basil Blackwell.

Horelli, L. (2017). Engendering urban planning in different contexts. *European Planning Studies*, 25(10), 1779–1796.

Hudson, C., & Rönnblom, M. (2020). Is an 'other' city possible. *Futures*, 121, Article 102583.

Iannelli, L., & Musaro, P. (Eds.). (2017). *Performative citizenship. Public art, Urban design, and political participation*. Mimesis International.

Jang, S.G., & Gim, T.H.T. (2022). Considerations for encouraging citizen participation by information-disadvantaged groups in smart cities. *Sustainable Cities and Society*, 76, 103437.

Jungk, R., & Mullert, N.R. (1998). *Håndbog i fremtidsværksteder*. Politisk revy.

Kern, L. (2020). *Feminist city*. Verso.

Kesselring, S., & Freudendal-Pedersen, M. (2021). Searching for urban mobilities futures: Methodological innovation in the light of COVID-19. *Sustainable Cities and Society*, 75, 103138.

Khakee, A. (2010). Futures studies and strategic planning. In M. Cerreta, G. Concilio, & V. Monno (Eds.), *Making strategies in spatial planning: Knowledge and values* (pp. 209–220). Springer.

Kirkland, E. (2008). What's race got to do with it? Looking for the racial dimensions of gentrification. *The Western Journal of Black Studies*, 32(2), 18–30.

Kirklies, P.C., Neumann, O., & Hohensinn, L. (2024). Promoting digital equality in co-production: The role of platform design. *Government Information Quarterly*, 41(1), 101903.

Kitchin, R., Lauriault, T.P., & McArdle, G. (Eds.). (2018). *Data and the city*. Routledge.

Kitchin, R., & Lysaght, K. (2003). Heterosexism and the geographies of everyday life in Belfast, Northern Ireland. *Environment and Planning*, 35, 489–510.

Lefebvre, H. (1968). *Le droit à la ville*. Seuil.

Legacy, C. (2017). Is there a crisis of participatory planning? *Planning Theory*, 16(4), 425–442.

Marvin, S., & Luque-Ayala, A. (2017). Urban operating systems. *International Journal of Urban and Regional Research*, 41(1), 84–103.

McAuliffe, C., & Rogers, D. (2018). Tracing resident antagonisms in urban development: Agonistic pluralism and participatory planning. *Geographical Research*, 56(2), 219–229.

Monno, V., & Khakee, A. (2012). Tokenism or political activism? *International Planning Studies*, 17(1), 85–101.

Mosciaro, M. (2021). Selling Milan in pieces: The finance-led production of urban spaces. *European Planning Studies*, 29(1), 201–218.

Mosco, V. (2019). *The smart city in a digital world*. Emerald.

Mullick, M., & Patnaik, A. (2022). Pandemic management, citizens and the Indian smart cities: Reflections from the right to the smart city and the digital divide. *City, Culture & Society*. doi:10.1016/j.ccs.2022.100474

Offenhuber, D. (2019). The platform and the bricoleur. *Urban Analytics and City Science*, 46(8), 1565–1580.

Parent, L. (2018). *Rouler/Wheeling Montréal: Moving through, resisting and belonging in an ableist city* [PhD thesis]. Concordia University.

Paton, K. (2014). *Gentrification: A working-class perspective*. Ashgate.

Peck, J. (2005). Struggling with the creative class. *International Journal of Urban and Regional Research*, 29(4), 740–770.

Pluchinotta, I., Esposito, D., & Camarda, D. (2019). Fuzzy cognitive mapping to support multi-agent decisions in development of urban policymaking. *Sustainable Cities and Society*, 46, 101402.

Purcell, M. (2008). *Recapturing democracy: Neoliberalization and the struggle for alternative urban futures* (1st ed.). Routledge.

Purcell, M. (2016). For democracy: Planning and publics without the state. *Planning Theory*, 15(4), 386–410.

Rogan, K. (2019). The universal factory. *ENQUIRY: The ARCC Journal for Architectural Research*, 16(2), 18–31.

Rose, G. (2020). Actually-existing sociality in a smart city. *City*, 24(3–4), 512–529.

Savini, F., & Aalbers, M.B. (2016). The de-contextualisation of land use planning through financialisation: Urban redevelopment in Milan. *European Urban and Regional Studies*, 23(4), 878–894.

Schmid, C. (2012). Henri Lefebvre, the right to the city, and the new metropolitan mainstream. In N. Brenner, P. Marcuse, & M. Mayer (Eds.), *Cities for people, not for profit: Critical urban theory and the right to the city* (pp. 42–62). Routledge.

Schwencke, E. (2017). Kritisk utopisk aksjonsforskning (CUAR) og utfordringer i deltakende prosesser. In S. Gjøtterud, H. Hiim, D. Husebø, L.H. Jensen. T. Steen-Olsen, & E. Stjernstrøm (Eds.), *Aksjonsforskning i Norge. Teoretisk og empirisk mangfold* (pp. 357–378). Cappelen Damm.

Shelton, T., & Lodato, T. (2019). Actually existing smart citizens. *City*, 23(1), 35–52.

Späth, P., & Knieling, J. (2018). Endlich smart-city-Leuchtturm. InA. Bauriedl & A. Strüver (Eds.), *Smart city – Kritische Perspektiven auf die Digitalisierung in Städten* (pp. 345–356). Transcript.

Stavanger Municipality. (2023). *Beautiful. Sustainable. Together.* NEB-STAR. https://nebstar.eu/

Umemoto, K. (2001). Walking in another's shoes: Epistemological challenges in participatory planning. *Journal of Planning Education & Research*, 21, 17–31.

United Nations. (2017). *New urban agenda* (Resolution 71/256).

Utt, J. (2021). Complicating the ally/enemy dichotomy: White teachers, critical Whiteness, and racial justice identifications. In C. Hayes, I.M. Carter, & K. Elderson (Eds.), *Unhooking from Whiteness: It's a process* (pp. 180–202). Brill.

Vanolo, A. (2014). Smartmentality. *Urban Studies*, 51(5), 883–898.

Vasudevan, R. (2020). The potential and pitfalls of 'art in research' methodologies. *Planning Theory & Practice*, 21(1), 58–75.

Waldron, R. (2019). Financialization, urban governance and the planning system: Utilizing 'development viability' as a policy narrative for the liberalization of Ireland's post-crash planning system. *International Journal of Urban and Regional Research*, 42(4), 685–704.

Westenberg, P., & Rutten, K. (2017). 'Do you speak my neighbourhood?' Language, technology, and proximity. *Critical Arts*, 32(2), 110–126.

Williams, M., & Sharif, N. (2021). Racial allyship: Novel measurement and new insights. *New Ideas in Psychology*, 62, 100865.

Willis, K.S. (2019). Whose right to the smart city? In P. Cardullo, C. Di Feliciantonio, & R. Kitchin (Eds.), *The right to the smart city* (pp. 27–41). Emerald.

Wood, D.M., & Mackinnon, D. (2019). Partial platforms and oligoptc surveillance in the smart city. *Surveillance & Society*, 17(1/2), 176–182.

7 Conceptual barriers to integrating smart and sustainable mobility planning

Daniela Müller-Eie

Introduction

While the *smart city* concept has been ubiquitous in urban planning practice for the last decades, urban planners have not actively engaged with and championed the potential of the smart city (Karvonen et al., 2020). When scrutinizing the smart city theoretically and conceptually, it becomes challenging to integrate smart approaches into urban planning theory and practice without reservations. To use urban mobility as an example, it is difficult to reconcile smart initiatives aiming at making private car travel easier or more convenient by ways of automated and autonomous vehicles or intelligent route choice and parking systems, and policies and research attempt to find ways to effectively curb the growth in private motorized urban travel by improving public and active travel as well as restricting and reducing the attractiveness of car use. Given the precarity of achieving more sustainable mobility in cities in terms of environmental impact, traffic safety, transport equity, and urban environmental quality, the contribution of smart approaches, and whether they are the best way forward, needs to be questioned critically.

To address the integration of the smart city into urban mobility planning, this conceptual chapter explores the nature of the *smart city* concept and its theoretical grounding and practical application in terms of epistemology and instrumentality. Barriers to integrating smart approaches in current research and practice for sustainable urban mobility are uncovered.

Firstly, it is problematic that the *smart city* concept lacks a clear definition and operationalization. With a concept as weakly defined and all-compassing as the smart city, it is difficult to link it to existing urban planning theory. Within the existing conceptualizations, it seems the smart city is more explorative and experimental in nature than the inherently normative discipline of planning. Second, it is unclear whether smart mobility measures are viewed and implemented as an end in themself (i.e., to make urban mobility smarter) or as a means to an end (i.e., using smart approaches to achieve something else). This is an essential distinction in urban planning. And, if the smart city is a means to an end, it is unclear which goal is to be achieved. Efficiency, simplicity, digital integration, or sustainability? Third, if smart mobility is viewed as a means to the end of

DOI: 10.4324/9781003498650-10

sustainable urban mobility, then it must be compared to other means. For now, it is uncertain to which extent smart approaches contribute to making urban mobility more sustainable and, therefore, how much focus they should receive compared to other, maybe more conventional, mobility planning measures. This is particularly important when resources are limited and focus on smart mobility can deter alternative approaches.

The chapter concludes with a summary of barriers to integrating smart and sustainable mobility and a call for further scrutiny when implementing smart city approaches in urban mobility planning.

The smart city concept

Researchers have tried to define and operationalize the *smart city* concept more concretely (Caragliu et al., 2011). Smart city concepts often relate to the use of technology to make certain elements of the city (e.g., administration, education, healthcare, public safety, real estate, transportation, utilities) 'more intelligent, interconnected, and efficient' (Brown et al., 2020, p. 3; Kitchin, 2015). Thus, they can be characterized by the application of information and communications technology/Internet of Things, competition, open data, public–private collaboration, and user involvement. Important stakeholders in the smart city are the urban government, planners, politicians, technology companies, knowledge organizations, and inhabitants (Jakobsen, 2019). Albino et al. (2015) reviewed smart city definitions and dimensions in an in-depth literature review and found that definitions of what constitutes a smart city lack universality. Therefore, they called for a common definition that integrates both *hard* (i.e., high-tech) and *soft* (i.e., non-tech) domains (Albino et al., 2015; Mora et al., 2017), at the same time as they advocate local contextualization of 'smartness' and its assessment.

While the concept and its applications have become more concrete in recent decades, they have also become broader. The smart city now comprises almost all aspects of urban structure and life, growing from mere technological applications to including governance and democracy, education and participation, as well as smart art (see Alberts and Fadnes, chapter 9). The word *smart* is also used differently in different fields, and related concepts can be technological, human, or institutional. In technology and science, smart denotes intelligent-acting products, AI, machine thinking/learning, automatic self-computing, or connected sensors. In planning, it can refer to a normative, ideological, or strategic direction that aims at optimizing resources (Neirotti et al., 2014) linked to the improvement of quality of life (Giffinger & Gudrun, 2010; Neirotti et al., 2014). In marketing, it demarcates greater value from the user perspective (Nam & Pardo, 2011).

Thus, the *smart city* concept seems to be defined through the application of technology and hard measures on one hand and through economic competitiveness and human prosperity and soft approaches on the other. The fact that it is applied to many facets of cities and urban life means that many components are included in one concept. While this makes the *smart city* concept more holistic, it also leads to more complexity. Applying the *smart city* concept within many disciplines leads to a proliferation of understandings (Albino et al., 2015). This way, the smart city has become the ultimate enigmatic canvas, onto which different fields can project meaning. The fact that the smart city remains 'quite fuzzy' (Caragliu et al., 2011, p. 67) and a 'diffuse and diverse entity' (Brown et al., 2020, p. 4) contributes to it being difficult to define and operationalize, while it allows it to be filled with a multitude of meanings. Therefore, there are still calls for a clearer and commonly accepted definition of the concept (Brown et al., 2020; Debnath et al., 2014). This conceptual diffusion is a general challenge and becomes even more challenging when integrating the smart city concept in the field of urban planning.

Urban planning and the smart city

When viewing the *smart city* concept through the lens of urban planning, more fundamental discussions about the concept's epistemology and instrumentality emerge. Urban planning is mainly concerned with developing urban physical and policy frameworks that improve physical urban environments and urban life. While norms and goals about what constitutes such an improvement have changed over time, the last decades have been thoroughly informed by a strong normative pressure to achieve environmental, social, and economic sustainability; resilience; social justice; and quality of life (Kelley et al., 2020). Thus, urban planning often follows relatively comprehensive approaches to mapping and analyzing urban conditions and challenges, identifying possible solutions and improvements, and implementing measures aiming at urban sustainability.

In terms of mobility planning, this translates into the objectives of reducing greenhouse gas emissions, reducing private car use, increasing public and active urban travel, increasing traffic safety, distributing access to facilities and transport equitably, increasing public health, and improving the quality of the urban environments and life. To achieve these objectives, several means are commonly proposed and implemented, such as strategic land use planning (e.g., densification, transit-oriented development, mixed use, reduced parking), improvement of infrastructure for public and active transport (e.g., light rail, bus rapid transit, cycle highways), and financial incentives (e.g., road tolls, parking fees, discounted transit fares). Thus, in urban planning and specifically in mobility planning, there are typically clearly assumed causal relationships between a certain goal that is to be achieved, a local context, and a measure or intervention that can be implemented to achieve the goal. This way of planning can be viewed as rational, contextual, and normative.

The *smart city* concept, on the other hand, is often characterized by technological advancement and testing and, as such, is more explorative and experimental in nature (Cowley & Caprotti, 2019; Cugurullo, 2018; Mora et al., 2017). Also, where smart initiatives are concerned with deploying new technology and proving that it works, urban planners must critically assess the cost-effectiveness, longevity, and political feasibility of measures. As such, smart city projects may be less familiar with day-to-day planning practices and alternative solutions. These intrinsic differences may cause difficulties when integrating smart city approaches in urban mobility planning.

Lack of contextuality, comprehensiveness, and norms in the smart city

When exploring the overlap between urban planning and the smart city, it seems obvious that urban planning should play a key role in grounding the *smart city* concept through implementation into the physical urban realities (Papa et al., 2013). However, exploring smart city projects, practices, and politics, Karvonen et al. (2020, p. 65) found that 'the influence of urban planners had been surprisingly muted' and called for a more active engagement of urban planners with the smart city. A reason for the lack of engagement may be a discrepancy in where and how the smart city measures are implemented. While the concept per definition concerns urban environments and urban life, it does not necessarily constitute or relate to physical realities. When the concept refers to the implementation of technology, participatory governance, or the use of open data (Komninos et al., 2019), such smart approaches are often non-spatial or non-contextual and readily applied and translated globally. This is contradictory to the prevalent notion in urban planning that approaches should be context-specific and adapted to local conditions. Thus, there seems to be a dichotomy between the hypothesized and imagined smart city and the physical and lived smart city (Aurigi & Odendaal, 2021).

Moreover, while some argue that full urban-scale smart city approaches are nothing but comprehensive rational planning, others characterize the smart city as 'anti-planning' (Cowley & Caprotti, 2019, p. 430) or 'planning without a plan' (Komninos et al., 2019, p. 6). This represents the perspective that smart city approaches are somehow opposed to and replacing the normative foundations of urban planning 'with notions of efficiency, standardization, and corporate control' (Karvonen et al., 2020, p. 66). They also seem to be 'shaped by a series of opportunities' rather than strategically thought-out plans (Komninos et al., 2019, p. 15), which undermines the rationale behind the existence of planning. An example is the introduction of five electric buses in Stavanger between 2015 and 2017 as part of the Triangulum project. Since this pilot project, the local bus fleet has continued to be run on fossil fuel because the pilot was not integrated into a larger comprehensive plan. This 'making of city plans comes closer to the concept of a laboratory of ideas and a roadmap of open innovation and entrepreneurship [...] than meticulous elaboration and implementation of plans by central and local authorities' (Komninos et al., 2019, p. 16). The 'experimental qualities of the smart city [...] issue a challenge to the very ambition of "planning" the urban future' (Cowley & Caprotti, 2019, p. 441). The matter of a trial-and-error approach, testing and piloting technologies, and the related techno-optimism may be what is intriguing about the *smart city* concept to some (Cowley & Caprotti, 2019) but is generally not in line with fundamental urban planning ideals and practice.

There is also a general concern that smart city approaches represent a move from normative and goal-driven planning towards more explorative and possibility-driven facilitation or, as Späth and Knieling (2020, p. 12) put it, 'a shift away from planning for a slowly transforming mobility system towards the management of flows "in real-time"'. When autonomous buses were introduced in Stavanger in 2022, they were reasoned to make public transport more attractive and address the first- and last-mile challenges. However, this is done without scrutiny or analysis of the actual need for or consequences of autonomous public transport, route choice, or user experiences. Thus, the introduction of autonomous buses lacks legitimacy in the normative construct that is the management of travel demand and supply. Hence, the smart city may constitute a 'fundamental shift in how planners approach the assessment of future needs in the city' (Späth & Knieling, 2020, p. 14).

The lack of local physical contextuality and coherent rationality to achieve normative goals in the *smart city* concept may be what prevents urban planners from fully embracing and integrating it. While the view of planning being rational and normative may be somewhat outdated or naive, any form of planning aims proactively at increasing efficiency, sustainability, or liveability through applying certain tools or processes. Merely standing by, observing, and analyzing urban form and function are not inherent to the nature of planning.

Top-down or bottom-up

Then why are smart city approaches heavily implemented? There seem to be two narratives of the smart city as either a bottom-up initiative or a top-down approach. Alkış et al. (2019, p. 16) wrote that 'the *smart city* vision arises from the necessity of management, automation, optimization, and exploration of all aspects of a city for improvement purposes', and they stated that 'citizens are forcing governments' to apply smart measures to improve resource efficiency; that is, portraying the concept as a bottom-up initiative. However, Späth and Knieling (2020) stated that the smart agenda is driven by technological research and development, governments, and funding bodies. This gives rise to the question of who has ownership of the smart city, who is served by the smart city, and how benefits and costs are distributed among those exposed to the smart city.

Local planning authorities and private investors, for instance, have adapted ongoing projects or existing plans, fitting them into a smart city framework (Haarstad & Wathne, 2018), thus harbouring benefits, momentum, and external funding that come with it. It can be hypothesized that this 'smart-washing' happens because urban planners are pressured to incorporate current objectives into a smart city framework to 'sell' their projects and receive funding. Haarstad and Wathne (2018) explained how such strategic application of the *smart city* concept to existing plans can result in projects being implemented quicker and increase trans-sectoral collaboration and the general level of ambitions. At the same time, most planners understand that focusing on technologically driven transport development is not sufficient to make urban mobility more sustainable and utilize the fact that the concept is not very well defined. This is how new bicycle infrastructure or city bike shares become part of smart mobility strategies when in reality they are conventional mobility planning. As such, even though the *smart city* concept is readily applied by urban planners, the cause may be more pragmatic and strategic than idealistic and evolving through bottom-up initiatives.

The smart city – means or end?

To integrate the *smart city* concept with the rational and normative approaches of urban planning, this chapter discusses two distinct ways of viewing the smart city: (a) as an *end in itself*, viewing the *smart city* as a conceptual goal to be achieved, or (b) as a *means to an end*, viewing smart approaches as a means to achieve other goals. The latter case leads to the questions of which goals and whether the *smart city* concept is the most appropriate way of achieving them.

Parts of the smart city discourse treat the concept as an end in itself, or at least without explicit reference to a goal. That is, when smart technologies are implemented, a higher degree of participation and co-creation is used and the city is viewed as a testbed to explore new technologies, business, and community models, which will make the city smarter. Viewing the concept as such, the achievement of the goal (i.e., smartness) is characterized through the application of its means (i.e., technology, participation, co-creation, etc.) This makes the concept a self-fulfilling and tautological construct.

Much of the smart city discourse, however, describes the concept as a means to several ends (Brown et al., 2020; Höjer & Wangel, 2015; Kelley et al., 2020; Martin et al., 2018; Soe, 2020). That is, when the city becomes smarter (using smart technologies, participation, co-creation, etc.), the city will be more sustainable, safer, and more resilient, equitable, healthy, or liveable. These possible goals can be continuously redefined through values and goals provided by urban communities and society at large. Supporting the view of smart approaches as a means to an end raises the question of the goal.

Smart to achieve sustainability

Soe (2020, p. 28) stated that '[…] the aim of cities pursuing to become smart should not be technology-driven but should help to solve actual global challenges on the broader level', and Komninos et al. (2019, p. 17) stated that 'evolutionary smart city planning goes far beyond the physical space of cities, addressing all the grand challenges of twenty-first-century life in cities', including growth, employment, poverty, sustainability, and safety, thus encompassing an array of economic, environmental, and social goals. Elsewhere, making cities more efficient is equated with making them more sustainable (Brown et al., 2020). Often, smart technology is applied to measure urban quality or behaviour (Ahlers et al., 2018) and cater to better decision making. In this sense, it does not solve anything but is a means to measure environmental impact. Achieving urban sustainability, and particularly Sustainable Development Goal 11, is frequently said to be the explicit aim of making cities smarter (Visvizi & Pérez del Hoyo, 2021). Haarstad and Wathne (2018, p. 2) characterized 'cities as strategic actors that utilize the resources from smart city discourses to mobilize initiatives, projects, and networks to benefit their priorities and interests – including sustainability ambitions'.

Smart mobility approaches, for instance, apply technological solutions, such as intelligent transport systems, to reduce driving and congestion and increase availability of parking spaces. Projects that integrate different travel modes or ticketing can make travel chains more efficient and seamless. Technology that supports shared mobility solutions, such as cars, city bikes, or e-scooters, reduces the need for ownership, gives access to a greater range of travellers, and addresses first- and last-mile issues. Thus, smart mobility projects are often framed as a means to urban sustainability and resilience (Kelley et al., 2020).

Being explicit about viewing the smart city as a means to achieving urban sustainability and mobility holds several benefits. First, it makes it easier to integrate the concept into existing planning approaches, viewing the smart city as a tool. Second, it provides grounds for evaluating the success or failure of a smart initiative when measured against its impact on environmental, social, or economic sustainability, thus providing a higher degree of accountability. However, this also requires a disentangling of the concepts of *smart* and *sustainable*.

Smart = sustainable?

The terms smart and sustainable are often used interchangeably without distinction between means and ends. This can become both confusing and challenging.

Cowley and Caprotti (2019) described how *smart* is discursively interwoven with *green* or *sustainable* from a conceptual point of view, and the smart city narrative seems closely linked to the concept of sustainability (Aurigi & Odendaal, 2021). Generally, there are several different and partly overlapping and competing urban concepts, but de Jong et al. (2015) encouraged rigour and nuance when using the terms smart and sustainable in the genealogy of the concepts. Others seem to promote a clear dichotomy between smart and sustainable, stating that 'the *smart city* competes with (and sometimes overshadows) sustainable urban development agendas' (Karvonen et al., 2020, p. 66). Jakobsen (2019) added a third dimension of urban ideals to this mix by stating that there are three competing visions of cities: sustainable, resilient, and smart. Elsewhere it was proposed that the concepts can be reconciled through smart sustainable, spatial, and human approaches (Roggema, 2020). The conceptual overlap is further evidenced by the introduction and frequent use of 'smart sustainable' (Bibri & Krogstie, 2017; Höjer & Wangel, 2015; see also Müller, chapter 2), with Huovila et al. (2019) calling for a simultaneous focus on both as if they were equal concepts, contradicting the definition of smart as a means to sustainability.

This calls for a better definition and clearer distinction between the terminology and the concepts of smart and sustainable, as the lack of definition and the all-encompassing nature of both concepts get in the way of clarity. It is also important to further scrutinize the relationship between the concepts. They could be viewed as equally important goals. Then the question is, how do they differ, and how can we prioritize between them? Or they could be seen as entirely interchangeable and overlapping. If so, then why do we need both? Or, as is argued in this text, one is a means to the other. When proposing to treat the *smart city* concept as a means to urban sustainability, this finally raises the question of whether smart approaches are the most effective means to achieve sustainable urban mobility.

Conventional sustainable urban mobility planning vs. smart mobility

While the *smart city* concept does impact the physical, social, economic, and political structure of cities, 'its net contribution to the environmental cause remains disputed' (de Jong et al., 2015, p. 34).

For sustainable urban mobility, one main objective is to reduce the negative environmental impact of car use. There are several steps to achieving this: (a) reduce the need and desire to travel by car; that is, disincentivize and decrease car ownership or offer work from home; (b) decrease travel distances through high-density mixed-use development with high-intensity of functions and facilities; (c) facilitate the modal shift from the car towards public or active travel; and (d) reduce emissions from motorized transport by increasing car occupancy and promoting low-emission vehicles running on renewable energy. These conventional approaches to sustainable urban transport require both infrastructural, policy, and behavioural changes.

Smart mobility, on the other hand, often refers to autonomous, intelligent, and cooperative vehicle technology (Maldonado Silveira Alonso Munhoz et al., 2020), intelligent fleet management, route finding, and parking administration (Satyakrishna & Sagar, 2018), smart ticketing systems, or electric vehicles and related charging infrastructure (Brown et al., 2020). Often simplifying the user experience is at the forefront. Not all of these measures align with conventional urban sustainable mobility planning. On the contrary, one could argue that, at least on the surface, technology catering to perpetuating private or motorized travel hinders existing sustainable urban mobility approaches. Even more disconcerting, it has been shown that the development and implementation of autonomous vehicles have led to users travelling 'more often and substitute trips previously made by transit, bicycling, or walking' (Kelley et al., 2020, p. 548).

Thus, when smart mobility approaches favour research and development of automated vehicles, car sharing or pooling, drones, or micro-mobility that perpetuates private car use or leads to substitution of public and active travel, they are counterproductive to existing efforts of reducing the need for travel and making urban mobility more sustainable. Späth and Knieling (2020) questioned whether focusing on technological solutions, efficiency, and related stakeholders distracts from conventional urban planning concepts substantiating efforts of more compact urban living, less need for travel, more public and active travel, and sustainable lifestyles. That is, are we addressing big societal challenges with technology because it is easier and allows us to perpetuate driving or consuming instead of acknowledging the need for fundamental change in urban structures and lifestyles? If so, 'the smart city policy discourse can become a smokescreen that distracts from the more immediate issues facing a city' (Aurigi & Odendaal, 2021, p. 58).

Elsewhere, definitions of smart mobility are more aligned with conventional strategies; that is, about changing travel habits, active modes of transport, cross-sector collaboration, citizen participation and empowerment, testing and evaluation, and in general transforming the mindset from traffic to mobility (Aarhus Kommune et al., 2017). This definition may be symptomatic of what the 'pragmatic and strategic manoeuvring within the relatively wide opportunity space that the "smart city" label affords' (Haarstad & Wathne, 2018, p. 20); that is, the re-branding of conventional approaches to sustainable urban mobility into smart mobility measures and aiming at objectives defined before and outside of the smart city agenda. Here, the previously described broadness and fuzziness of the *smart city* concept is an advantage. While Haarstad and Wathne (2018) optimistically reported on higher feasibility and higher levels of ambition in projects adapted to the smart city framework in this way, Späth and Knieling (2020) described smart city projects as only improving minorly on the status quo and lacking ambitions. This way, the *smart city* concept can contribute to complacency with the status quo and give certain groups of urban stakeholders the opportunity to test and pilot approaches without them leading to spatial, social, or environmental transformation.

How sustainable is smart mobility?

Previous studies showed that smart mobility measures are difficult to empirically assess regarding goal achievement and their contribution to environmental and social sustainability (Müller-Eie & Kosmidis, 2023). It is therefore fair to question whether efforts in smart mobility contribute to achieving sustainable mobility and whether they are the most effective way. Smart mobility approaches should be related to sustainable mobility and explicitly state how. Further, smart mobility projects (as, in fact, all mobility projects) must be assessed in terms of their contribution to sustainable, safe, and just mobility. That is, do they contribute to reducing the amount and distance of motorized travel, reducing environmental impact, increasing transport safety and equity, social acceptance, and quality of urban life? Additionally, the added value of smart mobility measures compared to conventional approaches should be evaluated. In this way, it becomes possible to assess whether smart mobility measures are effective and competitive means to foster urban sustainable mobility.

Conclusion

The *smart city* concept is often presented as a way to address the challenges that cities and urban areas face in terms of environmental stress, economic pressure, and social instability. This chapter critically reviewed the *smart city* concept and compared the inherent goals and objectives of smart approaches with conventional urban planning approaches within urban sustainability through examples of mobility. The findings can be summarized as follow:

1   The *smart city* concept is still lacking a strong theoretical grounding as a stand-alone concept as well as with regard to its role within the field of urban planning. On one hand, this makes it difficult to justify, implement, and evaluate smart approaches. On the other hand, this allows for a plethora of meanings to be projected onto the *smart city* concept, which may render it meaningless.

2   There are inherent conceptual difficulties in integrating the explorative, experimental, and pragmatic concept of smart cities with the inherently normative, rational, and idealistic field of urban (mobility) planning. These fundamental conceptual differences can lead to a rejection of smart approaches as being too descriptive and lacking clear goals.

3   Viewing the smart city as an end in itself to make cities more efficient becomes a tautological construct, achieving its goal by merely being implemented.

4   There are benefits to viewing the *smart city* concept as a means to achieving urban sustainability. This allows the concept to be more readily integrated into ongoing planning activity and leads to a higher degree of measurability and accountability. However, it is important to clarify how smart approaches differ from conventional approaches and what the added value is.

5   There is a need to explicitly communicate the impact of smart approaches on sustainable urban development. This requires a clear definition and operationalization of the approaches and a way to measure and assess their effect. To this end, smart urban mobility measures should be subjected to rigorous evaluation in terms of goal achievement.

6   There is a need for mapping the relationship between conventional sustainable urban mobility approaches and those applying smart technology to assess which are more effective and acceptable as it is possible that focusing on technological approaches distracts from efforts to replace car use by public or active travel. This is particularly important in practice where different projects need to be prioritized and implemented based on cost-efficiency.

Given the presented discussion and findings, it is not surprising that the *smart city* concept has been met with reluctance and ambiguity by urban planners. As shown, it will take some effort to conceptually reconcile smart and sustainable urban mobility, and it is only recommended to do so if smart approaches hold added value for achieving more sustainable urban travel.

Having said that, the *smart city* concept cannot and should not be ignored by urban planning for several reasons. Firstly, it has become an undeniable strategic force, financially incentivized by the European Union and other governments (Brown et al., 2020). Secondly, technological advances and digitalization have a profound impact on urban planning and may well hold solutions to some of the biggest challenges cities are facing in the future.

References

Aarhus Kommune, Trafik-, Bygge og Boligstyrelsen. (2017). *Smart mobilitet: Prosjektkatalog og evaluering*.

Ahlers, D., Kraemer, F.A., Bråten, A.E., Liu, X., Anthonisen, F.V., Driscoll, P.A., & Krogstie, J. (2018). Analysis and visualization of urban emission measurements in smart cities. *Advances in Database Technology – EDBT 2018*, 1–4. https://openproceedings.org/2018/conf/edbt/paper-339.pdf

Albino, V., Berardi, U., & Dangelico, R.M. (2015). Smart cities: Definitions, dimensions, performance, and initiatives. *Journal of Urban Technology*, 22(1), 3–21.

Alkış, N., Çaldağ, M.T., & Gökalp, E. (2019). An integrated holistic success model for evaluating smart city initiatives. In H.R. Arabnia, L. Deligiannidis, & F.G. Tinetti (Eds.), *Proceedings on the International Conference on Internet Computing (ICOMP)* (pp. 16–21). CSREA Press.

Aurigi, A., & Odendaal, N. (2021). From 'smart in the box' to 'smart in the city': Rethinking the socially sustainable smart city in context. *Journal of Urban Technology*, 28 (1–2),55–70.

Bibri, S.E., & Krogstie, J. (2017). Smart sustainable cities of the future: An extensive interdisciplinary literature review. *Sustainable Cities and Society*, 31, 183–212.

Brown, W., King, M., & Goh, Y.M. (2020). UK smart cities present and future: An analysis of British smart cities through current and emerging technologies and practices. *Emerald Open Research*, 2(4), 4.

Caragliu, A., Del Bo, C., & Nijkamp, P. (2011). Smart cities in Europe. *Journal of Urban Technology*, 18(2), 65–82.

Cowley, R., & Caprotti, F. (2019). Smart city as anti-planning in the UK. *Environment and Planning D: Society and Space*, 37(3), 428–448.

Cugurullo, F. (2018). Exposing smart cities and eco-cities: Frankenstein urbanism and the sustainability challenges of the experimental city. *Environment and Planning A: Economy and Space*, 50(1), 73–92.

Debnath, A.K., Chin, H.C., Haque, M.M., & Yuen, B. (2014). A methodological framework for benchmarking smart transport cities. *Cities*, 37, 47–56.

de Jong, M., Joss, S., Schraven, D., Zhan, C., & Weijnen, M. (2015). Sustainable–smart–resilient–low carbon–eco–knowledge cities; making sense of a multitude of concepts promoting sustainable urbanization. *Journal of Cleaner Production*, 109, 25–38.

Giffinger, R., & Gudrun, H. (2010). Smart cities ranking: An effective instrument for the positioning of the cities? *ACE: Architecture, City and Environment*, 4(12), 7–26.

Haarstad, H., & Wathne, M.W. (2018). Smart cities as strategic actors. In A. Karvonen, F. Cugurullo, & F. Caprotti (Eds.), *Inside smart cities: Place, politics and urban innovation* (pp. 102–116). Routledge.

Höjer, M., & Wangel, J. (2015). Smart sustainable cities: Definition and challenges. In *ICT innovations for sustainability* (pp. 333–349). Springer.

Huovila, A., Bosch, P., & Airaksinen, M. (2019). Comparative analysis of standardized indicators for smart sustainable cities: What indicators and standards to use and when? *Cities*, 89, 141–153.

Jakobsen, T.S. (2019). Er den Smart Byen bare et slagord? En kritisk guide. In I.M. Henriksen & A. Tjora (Eds.), *Bysamfunn* (pp. 225–235). Universitetsforlaget AS.

Karvonen, A., Cook, M., & Haarstad, H. (2020). Urban planning and the smart city: Projects, practices, and politics. *Urban Planning*, 5(1), 65–68.

Kelley, S.B., Lane, B.W., Stanley, B.W., Kane, K., Nielsen, E., & Strachan, S. (2020). Smart transportation for all? A typology of recent U.S. smart transportation projects in midsized cities. *Annals of the American Association of Geographers*, 110(2), 547–558.

Kitchin, R. (2015). Making sense of smart cities: Addressing present shortcomings. *Cambridge Journal of Regions, Economy, and Society*, 8(1), 131–136.

Komninos, N., Kakderi, C., Panori, A., & Tsarchopoulos, P. (2019). Smart city planning from an evolutionary perspective. *Journal of Urban Technology*, 26(2), 3–20.

Maldonado Silveira Alonso Munhoz, P.A., da Costa Dias, F., Kowal Chinelli, C., Azevedo Guedes, A.L., Neves dos Santos, J.A., da Silveira e Silva, W., & Pereira Soares, C.A.

(2020). Smart mobility: The main drivers for increasing the intelligence of urban mobility. *Sustainability*, 12(24), 10675.

Martin, C.J., Evans, J., & Karvonen, A. (2018). Smart and sustainable? Five tensions in the visions and practices of the smart-sustainable city in Europe and North America. *Technological Forecasting and Social Change*, 133, 269–278.

Mora, L., Bolici, R., & Deakin, M. (2017). The first two decades of smart-city research: A bibliometric analysis. *Journal of Urban Technology*, 24(1), 3–27.

Müller-Eie, D., & Kosmidis, I. (2023). Sustainable mobility in smart cities: A document study of mobility initiatives of mid-sized Nordic smart cities. *European Transport Research Review*, 15(1), 36.

Nam, T., & Pardo, T.A. (2011). Conceptualizing smart city with dimensions of technology, people, and institutions. In J. Bertot, K. Nahon, S. Chun, L. Luna-Reyes, & V. Atluri (Eds.), *Proceedings of the 12th Annual International Digital Government Research Conference: Digital government innovation in challenging times* (pp. 282–291).

Neirotti, P., De Marco, A., Cagliano, A.C., Mangano, G., & Scorrano, F. (2014). Current trends in smart city initiatives: Some stylised facts. *Cities*, 38, 25–36.

Papa, R., Gargiulo, C., & Galderisi, A. (2013). Towards an urban planners' perspective on smart city. *TeMA Journal of Land Use, Mobility and Environment*, 6(1), 5–17.

Roggema, R. (2020). Towards integration of smart and sustainable cities. In R. Roggema & A. Roggema (Eds.), *Smart and sustainable cities and buildings* (pp. 5–23). Springer International.

Satyakrishna, J., & Sagar, R.K. (2018, January 19–20). *Analysis of smart city transportation using IoT* [Paper presentation]. 2nd International Conference on Inventive Systems and Control (ICISC), Coimbatore, India.

Soe, R.-M. (2020). Mobility in smart cities: Will automated vehicles take it over? In *Smart governance for cities: Perspectives and experiences* (pp. 189–216). Springer.

Späth, P., & Knieling, J. (2020). How EU-funded smart city experiments influence modes of planning for mobility: Observations from Hamburg. *Urban Transformations*, 2(1), 1–17.

Visvizi, A., & Pérez del Hoyo, R. (2021). Sustainable Development Goals (SDGs) in the smart city: A tool or an approach? (An introduction). In A. Visvizi & R. Pérez del Hoyo (Eds.), *Smart cities and the UN SDGs* (pp. 1–11). Elsevier.

8 Addressing cyber-physical challenges for critical infrastructures in smart cities through integrating organizational processes for safety and security management

Ruth Østgaard Skotnes and Kenneth Pettersen Gould

Introduction

The concept of smart cities has been described as 'embodying the idea that information and communication technologies (ICTs) in urban development can lead to more efficient resource use, sustainable economic development, and a better quality of life' (König, 2021, p. 1). The global Sustainable Development Goals (United Nations, 2015) address cities and communities, especially with Goal 11, which will make them inclusive, safe, resilient, and sustainable. While 2030 is set as the deadline to achieve these goals, studies of their relationship to cyber-physical systems (CPS) are scarce (Broo et al., 2021). CPS are the confluence of ICT with physical infrastructure and systems (Humayed et al., 2017). CPS can be defined as 'integrations of computation, networking, and physical processes' (Lun et al., 2019, p. 174). New challenges for security and safety, caused by the emergence of CPS, lie at the intersection between various technologies and existing management scopes. Because of these developments, the traditional conceptual and organizational segregation (siloization) between cyber and physical safety and security professionals creates challenges for managing CPS risks. If cities are to realize the Sustainable Development Goals, ICTs must be designed and used in ways that minimize risks and avoid unwanted consequences, attacks, accidents, and disasters.

This chapter addresses social organization in relation to safety and security management by examining a particular aspect of smart city developments, namely, organizational challenges associated with the development and use of CPS in critical infrastructures such as electric power supply systems and water supply systems.

Critical infrastructures can be defined as the assets and systems that are necessary to maintain or restore a society's critical functions and fulfil the basic needs of society (DSB, 2016). Failures of critical infrastructures can represent a threat to people, to the economy, and to societal functions in a city, region, or country (Hokstad et al., 2012). Examples of cyber-physical technologies used in critical infrastructures are smart distribution and transmission, including advanced metering infrastructure in the electric power sector and smart water treatment and

DOI: 10.4324/9781003498650-11

distribution in the water supply sector (U.S. Department of Homeland Security, 2015). We have chosen to use these infrastructures as examples because they are two of the most critical infrastructures that increasingly use ICT systems for monitoring, control, and operation. The electric power system is a good example of technological interconnectivity: in a power outage, most services stop, and a prolonged interruption of power supply seriously affects other critical functions in society (Fridheim et al., 2001). Furthermore, the water and wastewater sector provides basic needs to both industry and citizens. Water is also both an input to hydroelectric plants and dependent on electricity for water treatment and distribution. Thus, these two sectors are both interconnected and vital for critical societal functions, with high requirements for both reliability and security (Hilt et al., 2018). The physical systems of electricity and water supply rely on the cyber systems and are thus CPS.

There has been an exponential growth in the development and deployment of various types of CPS providing the ability to monitor, manipulate, and automate devices. Many such systems are now integral to critical infrastructures and have become essential to our daily lives (Humayed et al., 2017). According to Wang et al. (2015), the term CPS was coined in the United States in 2006 (Lee, 2006), along with the realization of the increasing importance of the interactions between interconnected computing systems and the physical world. The term CPS is used as an umbrella term including robotics, machine automation, industrial control systems, the Internet of Things (IoT), and the Industrial Internet of Things (IIoT). All CPS have different applications, architectures, and behaviours when applied to critical infrastructures; however, they share some key attributes (Fink et al., 2018). CPS will usually include cloud platforms, embedded systems, and networks of sensors (Foch-Villaronga & Millard, 2019) that enable communication between the virtual and the physical.

CPS have been key targets in some of the most highly publicized security breaches over the last decade (Fink et al., 2018), generating increasing concerns about the interaction between digital devices and the existing physical equipment and operators in critical infrastructures. There has been a trend of cyberattacks against industrial control systems (process control systems) in the electric power supply sector, such as Stuxnet in 2010, BlackEnergy in 2015, Industroyer in 2016, and a large-scale cyberattack on Danish critical energy infrastructure in 2023 (Arghire, 2021, 2023; Perelman, 2018). Industrial control systems in water supply and wastewater treatment plants have also been attacked. Recent examples are the cyberattacks on several Israeli Water Authority facilities in 2020 (Olenick, 2020), a Florida city's water supply in 2021 (Greenburg, 2021), and reports that cyberattacks are hitting water and wastewater systems throughout the United States in 2024 (Lyngaas, 2024).

Security and safety

The human and organizational risks that arise in critical infrastructures from the development of CPS lie at the intersection between the domains of security and safety. *Security* refers to the risk of intended, malevolent events (e.g., cyberattacks or terrorist attacks), and *safety* usually refers to the risks and consequences of unintended, non-malevolent events (e.g., accidents or natural or man-made disasters), and the research fields of security and safety have until recently largely been treated separately (Glesner, 2022; Gould & Bieder, 2020). Similarly, the management of cyber security[1] and physical safety and security[2] has traditionally been performed as separate entities (silos) in public and private organizations. The information and cyber security management field (cyber domain) has been dominated by a focus on technological solutions, while the industrial safety tradition (physical domain) has developed a more holistic or integrated approach (including technical, human, and organizational aspects; Bartnes & Albrechtsen, 2016; Soomro et al., 2016).

According to Goslin (2016), there is a self-imposed segregation between cyber security and physical safety and security professionals. Additionally, there is often a further segregation between professionals responsible for physical safety (such as personnel safety or process safety) and physical security (such as access control to facilities; Ylönen et al., 2022). It is often argued that there are differences in knowledge, methods, mindsets, and risk perceptions between all these different professionals and that these differences may result in difficulties in coordinated cyber-physical threat mitigation (Johnsen, 2012). Through the development of CPS, the empirical distinctions between cyber security and physical safety and security are blurring. Combined with the self-imposed segregation between different safety and security professionals, the consequences are a lack of information flow and coordination, which creates gaps in protection (Goslin, 2016).

In this chapter, we attempt to broaden the perspectives of both the safety and security communities, advising them that management of critical infrastructures today depends on cross-sectoral and cross-departmental coordination, communication, cooperation, and knowledge sharing. A priority here concerns issues related to problems in the design of the interface between cyber security and physical safety and security. We will also address more extended concerns related to the question of why problems associated with system complexity and cascading effects across the cyber and physical domains are neglected.

Smart cities and regions are typically characterized by stakeholder heterogeneity, multi-layered authority structures, complex decision-making processes, competing objectives and goals, and networks of distinct legal entities with different technical infrastructures and capabilities (Bergan et al., 2020). Rasmussen's (1997) portrayal of complex sociotechnical systems in dynamic societies is constructed on such premises. Rasmussen argued for a system-oriented/holistic approach to societal risk management, an integrated approach across several levels ranging from legislators, via managers and work planners, to system operators – all in a context of rapid technological change, a competitive environment, and changing regulatory practices. According to Mondschein et al. (2021), the number of people in smart city and region governance responsible for managing integrative sociotechnical issues seem to be too few. We will suggest some steps that can be taken to change this.

In other words, this chapter looks closer at the organizational context that structures the relationship between the cyber system and the physical system of critical infrastructures in smart cities and regions, addressing how social organization is a key theme in relation to dealing with these concerns and that a holistic approach to safety and security management is necessary. The discussion in this chapter explores examples from Stavanger; however, it is relevant for smart city development in other regions and contexts as well.

Infrastructure management, city governance, and design

The development of CPS enables organizations that manage infrastructures to migrate controls from people to algorithm-based systems, thus increasing the efficiency of the systems while at the same time removing some of the human interactions within the system, thus mitigating potential for human errors. However, in addition to mitigating some risks, removal of the barrier between the cyber and physical worlds potentially introduces some new safety and security challenges. CPS are sensitive to a wider range of attacks and design flaws than traditional information technology (IT) systems and operational technology (OT) systems. Furthermore, the deployment of new smart city technologies introduces additional attack surfaces that malicious actors can target. These new cyber-derived threats are not fully covered by existing safety-oriented design methods, and the threat profile evolves rapidly. Examples of these new challenges are issues associated with increasing the number of system access points, and thereby the potential attack vectors; skill atrophy among human operators; loss of visibility into all parts of a system; necessary changes in emergency response plans (e.g., humans will not be present in areas of the system they once were); unanticipated permutations of automated functioning; or unintentional elimination of manual overrides (U.S. Department of Homeland Security, 2015). With smart technologies, failure in one system can have cascading effects that spread within and across systems, with potentially severe consequences for the reliability, safety, and security of critical infrastructures.

Traditionally, critical infrastructure safety and security was a physical matter, such as protecting a city's electricity supply against fire or hindering anyone from interfering with the water supply, as critical infrastructures operated based on mechanical or electronical devices in offline systems. However, these systems were expensive to deploy, maintain, and operate, and during the last decades they have been replaced or modified by new digital technologies (Kamp. 2016; Kriaa et al., 2015). *Industrial control system* is a general term that encompasses several types of control systems, including supervisory control and data acquisition (SCADA) systems and other process control system configurations, often found in cities' critical infrastructures (Stouffer et al., 2011). A SCADA system is a computer-based process control system used by national infrastructure utility systems, such as water supply, electric power generation and distribution, mass transportation, and oil and gas production and distribution. SCADA systems help control and monitor such utilities by gathering field data from sensors and instruments located at remote sites, transmitting, and displaying these data at a central site and enabling engineers to send control commands to field instruments; for example, track switches, gas and water pumps, traffic signals, valves, and electric circuit breakers (Patel & Sanyal, 2008). These types of control systems have become critical to the operation of highly connected and mutually dependent critical infrastructures (Stouffer et al., 2011). According to Patel and Sanyal (2008), a real-world SCADA system can analyze the data collected from hundreds to thousands of data collection points located in outdoor fields such as an airport, a gas distribution station, or an electric subdivision. The owners of these previously isolated systems have increasingly adopted ICT solutions to promote corporate business systems' connectivity and remote access capabilities, often connected to the internet, reducing the isolation of industrial control systems from the outside world (Leith & Piper, 2013). Consequently, safety and security of a critical infrastructure has become not only a solely physical practice but also a concern of securing digital/cyber space (Kamp, 2016).

Mondschein et al. (2021) found that several of the challenges and solutions to smart cities were organizational and managerial in nature. When infrastructure systems such as electric power supply and water supply are brought online, substantial expenses are required to integrate the old legacy systems with the modern systems. According to Mondschein et al., this is not just a technological issue but a coordination issue as well. In most cities around the world, historical asynchronization of legacy system purchases across different city departments has resulted in the presence of disparate systems that do not interface well and requires coordination among business partnerships and procurements across city departments. They found that new initiatives that required inter-department coordination faced significant challenges. Siloes of information were created that struggled to interact with each other.

Communication and coordination challenges

In critical infrastructures, communication and coordination challenges between organizational departments, units, and professions are, as previously mentioned, related to the increased interconnectivity of industrial control systems (Johnsen et al., 2009). The incorporation of ICT in industrial control systems results in increased vulnerabilities and threats and necessitates collaboration between departments, units, and professions that previously had non-overlapping or separate areas of responsibility. Generally, in critical infrastructure companies and other organizations that use similar technological systems, responsibility for seeing and securing that different groups of equipment designers and operators take account of each other's designs, vulnerabilities, and risks across the cyber and physical domains of a system has yet to be formalized in the position of a particular group of professionals.

As mentioned above, the structural segregation (*siloization*) between cyber and physical safety and security professionals becomes a problem and creates challenges for managing CPS risks. For instance, a study by Jaatun et al. (2009) identified the existence of both interdependencies and mistrust among IT personnel and traditional process control engineers using OT. Collaboration challenges among technical professionals and business staff were also reported (Ahmad et al., 2012). In 2019, a survey in the Norwegian water and electric power supply sectors (Malmedal & Skotnes, 2019) found that 24% of the respondents (from both sectors) answered that there was little communication and cooperation between the different departments/units/persons responsible for security and safety in their organizations. Moreover, many respondents answered 'Don't know' to this question (48% in the power sector and 38% in the water sector). Similarly, Ylönen et al. (2022) found that cyber security in high-risk process industries was often handled in a separate IT department, and the communication with the process safety department and the environment, health, safety, and security department was often inadequate. In addition, experts from different departments sometimes used similar terms that had different connotations in their respective fields or different terms and definitions that others did not understand. Thus, even when actors, groups, and departments responsible for safety and security management do interact, the cultural, educational, and experiential divides make effective communication between them a challenge (Gould & Bieder, 2020).

However, since increased digitalization and the implementation of smart city/region solutions create CPS, critical infrastructure organizations are required to look beyond these traditional boundaries and search for means of alignment between actors, groups, and departments responsible for safety and security management. Moreover, while cyber security traditionally has emphasized reliability and data access restrictions (Albrechtsen, 2015), smart city and region governance need to emphasize technology-enabled innovation by facilitating openness and data exchange across units and organizational boundaries. Thus, to manage risks, reduce vulnerabilities, and strengthen the resilience of critical infrastructures, different organizational actors need close collaboration. According to DeLoach (2015) and Renn et al. (2018), a siloed approach to managing risk is dangerous in today's rapidly changing environment, and a holistic approach to safety and security management is necessary.

Breaking down the silos

According to Wolf and Serpanos (2018), it is important for CPS to be designed and operated under a unified view of safety and security characteristics. The development of smart cities requires a new generation of professionals who not only are able to program automation algorithms but also have a better understanding of the societal implications of digitalization and automation. This will require managers and employees who are able to communicate and work within multidisciplinary groups from different knowledge silos, with different experiences, mindsets, and agendas (see also Huang et al., chapter 3). There is a need for scientists, engineers, planners, and risk, safety, and security managers who can comprehend and handle the risks, complexities, and uncertainties of CPS. They should be able to perform joint risk and vulnerability assessments and to communicate effectively across different multidisciplinary boundaries that do not traditionally share the same knowledge and strategic thinking. They must also be able to create appropriate ethical-by-design systems that accommodate future digitalized and automated transitions. The demand for such specialists with appropriate competences and skills needs to be acknowledged and advanced, in practice, as well as in research and education, to promote sustainable and future-oriented CPS (Broo et al., 2021).

Previous studies have suggested that managers' engagement and involvement are important for safety and security management (Antonsen, 2009; Hagen, 2009; Skotnes, 2015). If managers are engaged in this issue, they will be aware of the need for safety and security management and can ensure that measures are implemented. However, if managers want to prioritize holistic safety and security management of CPS, what can they do to break down the silos?

Glesner (2022), for instance, suggested creating 'tension venues' (p. 199). Safety and security share an overarching goal of protecting against dangers, but they deal with different types of hazards. Hence, while safety and security policies share some elements, their practical enactment may diverge and lead to potential tensions. Glesner thus proposed creating venues (e.g., regular meetings, common inspections, or common training courses) where different safety and security professionals discuss and raise tensions, share their stances and perspectives, find ways to overcome them, and find new solutions. Ylönen et al. (2022) also emphasized that it is necessary to think beyond the current disciplinary boundaries. They suggested establishing permanent forums where different experts can communicate together and co-construct a better understanding of convergent risks. Additionally, they suggested the introduction of new risk identification methods, with better integration of process safety, cyber security, and physical security risks and the co-assessment of these risks.

A possible solution regarding smart cities is the development of networks where actors from different silos can come together to share information and cooperate, with the goal of reducing cyber-physical safety and security challenges in smart cities and regions. To illustrate this, we will use an example from the city and municipality of Stavanger, which was the first municipality in Norway to adopt a local roadmap for the smart city in 2016 (Stavanger Municipality, 2016).[3] Municipalities in Norway have key responsibilities for safety and security of critical infrastructures within their geographic area. The governance structures of many of these infrastructures, however, are complex and involve organizations on municipal, regional, and national levels. This complexity is clearly visible in Stavanger, especially in the two sectors highlighted in this article: electric power and water. In addition to the municipality itself, there are several private companies responsible for managing critical infrastructures in the electric power and water sectors, which adds even more to the complexity. Yet, in Stavanger, a group of actors from the municipality, the university, Digi Rogaland,[4] and the smart city cluster Nordic Edge[5] have established a regional network for cyber security, where cyber-physical safety and security in the smart city/region is one of the focus areas. Several departments from the municipality are represented in the network, such as the innovation and digitalization department (which includes the former Smart City

Office), and representatives from the regional electric power and water companies are also invited to join the network. The network's mandate is to contribute to creating a safer and more secure society by adopting a holistic approach to safety and security management in the smart city/region; taking a leading role in addressing cross-sectoral challenges; facilitating joint learning and mutual sharing of information, experience, and expertise; contributing to more research on cyber security and cyber-physical challenges; and connecting the public and private sectors, academia, and citizens to raise the awareness of these challenges.

In this way, a regional or city network for cyber security can be one possible solution to start breaking down the silos and promote cross-sectoral coordination, communication, cooperation, and knowledge sharing among actors who are directly and/or indirectly responsible for the safety and security management of critical infrastructures within smart cities. In addition, public and private actors and companies on all structural levels need to work towards breaking down the silos within their own organizations. Managers should play a vital role when it comes to facilitating such a process. Lastly, academia plays an important role when it comes to developing and strengthening research and education on safety and security management of CPS, educating professionals who can comprehend and handle the risks, complexities, and uncertainties of CPS.

Concluding remarks

In this chapter we have taken a closer look at the organizational context that structures the relationship between the cyber system and the physical system of critical infrastructures in smart cities and regions, addressing how social organization is a key theme in relation to dealing with these concerns and that a holistic approach to safety and security management is necessary.

The possible risks involved in the transition towards smarter cities are related not only to a 'final state', after systems have become more digitalized, but also to a form of transition risk, related to the processes themselves. This transition involves risks of creating new blind spots in organizational memories, in the sense that both information and important tacit knowledge may be lost in translation on the way towards digitalized systems (La Porte, 2020). On the other hand, the transition provides opportunities to map work processes and explicate tacit safety and security knowledge if the change processes also involve thorough sociotechnical assessment of safety and security-critical success and failure factors.

In this chapter, we have focused on critical infrastructures in the electric power supply and water supply sectors. However, the challenges we highlighted will also be transferable to and relevant for other critical societal functions and infrastructures.

According to Cassandras (2016), the emerging prototype for a smart city is characterized by an urban environment with a new generation of innovative services for transportation, energy distribution, healthcare, environmental monitoring, business, commerce, emergency response, and social activities. Cassandras argued that this requires a viewpoint of smart cities as CPS that include new software platforms and strict requirements for mobility, security, safety, privacy, and the processing of massive amounts of information ('big data'). The data collected and flowing through such CPS may involve traffic conditions; the occupancy of parking spaces; air/water quality information; the structural health of bridges, roads, or buildings; and the location and status of city resources including transportation vehicles, police officers, or healthcare facilities. The applications stemming from these environments are virtually endless and are being invented and deployed daily.

Technology alone cannot transform a city without the participation and cooperation of its citizens towards future-oriented and sustainable solutions (Broo et al., 2021; Cassandras, 2016; United Nations, 2015). A smart sustainable city is a sociotechnical ecosystem of people, technology, organizations, and information. Cassandras (2016) emphasized that issues relating to smart cities can only be successfully addressed through a multidisciplinary approach, bringing together researchers from engineering, computer science, and the social sciences. This corresponds well with our argument that the continued development of smart sustainable cities will require employees who are able to communicate and work within multidisciplinary groups from different knowledge backgrounds, with different experiences, mindsets, and agendas. We will need scientists, engineers, planners, and risk, safety, and security managers who can comprehend and handle the risks, complexities, and uncertainties of CPS.

Notes

1 Cyber security is concerned with protecting 'everything' that is vulnerable because it is connected to or otherwise dependent on ICT (Norwegian Ministeries, 2019).
2 Securing the physical domain is the practice of protecting assets and people by physical means (Kamp, 2016).
3 Since 2019 the city has also followed the *Roadmap for Smart and Sustainable Cities and Communities in Norway* (DOGA et al., 2019).
4 Digi Rogaland is a collaboration in which the municipalities in Rogaland County have joined forces to ensure that the region's residents have equal access to digital tools and services. Stavanger Municipality is one of the members of Digi Rogaland. https://digir ogaland.no/
5 Nordic Edge is a non-profit organization working in close cooperation with private companies, municipalities, academia, and citizens towards smarter and more sustainable cities and communities. https://nordicedge.org/

References

Ahmad, A., Hadgkiss, J., & Ruighaver, A.B. (2012). Incident response teams – Challenges in supporting the organizational security function. *Computers & Security*, 31(5), 643–652. doi:10.1016/j.cose.2012.04.001

Albrechtsen, E. (2015). Major accident prevention and management of information systems security in technology-based work processes. *Journal of Loss Prevention in the Process Industries*, 36, 84–91. doi:10.1016/j.jlp.2015.05.004

Antonsen, S. (2009). *Safety culture: Theory, method and improvement*. Ashgate. doi:10.1201/9781315607498

Arghire, I. (2021, March 22). *Electricity distribution systems at increasing risk of cyber-attacks, GAO warns. Security Week.* https://www.securityweek.com/electricity-distribution-systems-increasing-risk-cyberattacks-gao-war/

Arghire, I. (2023, November 14). 22 Energy firms hacked in largest coordinated attack on Denmark's critical infrastructure. *Security Week.* https://www.securityweek.com/22-energy-firms-hacked-in-largest-coordinated-attack-on-denmarks-critical-infrastructure/

Bartnes, M., & Albrechtsen, E. (2016). Examining the suitability of industrial safety management approaches for information security incident management. *Information & Computer Security,* 24(1), 20–37. doi:10.1108/ICS-01-2015-0003

Bergan, P., Mölders, A.-M., Rehring, K., Ahlemann, F., Decker, S., & Reining, S. (2020). Towards designing effective governance regimes for smart city initiatives: The case of the city of Duisburg. In T.X. Bui (Ed.), *Proceedings of the 53rd Hawaii International Conference on System Sciences* (pp. 2323–2332). AIS eLibrary. doi:10.24251/HICSS.2020.283

Broo, D.G., Boman, U., & Törngren, M. (2021). Cyber-physical systems research and education in 2030: Scenarios and strategies. *Journal of Industrial Information Integration,* 21, 100192. doi:10.1016/j.jii.2020.100192

Cassandras, C.G. (2016). Smart cities as cyber-physical social systems. *Engineering,* 2, 156–158. doi:10.1016/J.ENG.2016.02.012

DeLoach, J. (2015, March 27). *Think holistically when managing risk.* Corporate Compliance Insights. https://www.corporatecomplianceinsights.com/think-holistically-when-managing-risk

DOGA, Norwegian Smart City Network, & Nordic Edge. (2019). *Roadmap for smart and sustainable cities and communities in Norway. A guide for local and regional authorities developed by Design and Architecture Norway (DOGA), the Norwegian Smart City Network and Nordic Edge.*

DSB. (2016). *Samfunnets kritiske funksjoner: Hvilken funksjonsevne må samfunnet opprettholde til enhver tid?* Versjon 1.0. Direktoratet for samfunnssikkerhet og beredskap. https://www.dsb.no/globalassets/dokumenter/rapporter/kiks-2_januar.pdf

Fink, G.A., Edgar, T.W., Rice, T.R., MacDonald, D.G., & Crawford, C.E. (2018). Overview of security and privacy in cyber-physical systems. In H. Song, G.A. Fink, & S. Jeschke (Ed.), *Security and privacy in cyber-physical systems: Foundations, principles, and applications* (pp. 1–24). John Wiley & Sons.

Foch-Villaronga, E., & Millard, C. (2019). Cloud robotics law and regulation – Challenges in the governance of complex and dynamic cyber–physical ecosystem. *Robotics and Autonomous Systems* 119, 77–91. doi:10.1016/j.robot.2019.06.003

Fridheim, H., Hagen, J., & Henriksen, S. (2001). *En sårbar kraftforsyning – Sluttrapport etter BAS3* [FFI/RAPPORT-2001/02381]. Forsvarets forskningsinstitutt. https://publications.ffi.no/en/item/asset/dspace:2962/03-01409.pdf

Glesner, C. (2022). *Safety and security in and through practice: Tensions at the interface* [Unpublished doctoral dissertation]. Université de Liège.

Goslin, C. (2016). *Eliminating the gaps between physical and cyber security.* Butchko esi. https://butchkoinc.com/eliminating-the-gaps-between-physical-and-cyber-security/

Gould, K.P., & Bieder, C. (2020). Safety and security: The challenges of bringing them together. In C. Bieder & K. Pettersen Gould (Eds.), *The coupling of safety and security* (pp. 1–8). Springer.

Greenburg, A. (2021, February 8). A hacker tried to poison a Florida city's water supply, officials say. *Wired.* https://www.wired.com/story/oldsmar-florida-water-utility-hack/.

Hagen, J.M. (2009). *The human factor behind the security perimeter – Evaluating the effectiveness of organizational information security measures and employees' contributions to security* [Unpublished doctoral dissertation]. University of Oslo.

Hilt, S., Huq, N., Kropotov, V., McArdle, R., Pernet, C., & Reyes, R. (2018). *Exposed and vulnerable critical infrastructure: Water and energy industries.* Trend Labs (The Global Technical Support and R&D Center of TREND MICRO). https://documents.trendm icro.com/assets/white_papers/wp-exposed-and-vulnerable-critical-infrastructure-the-wa ter-energy-industries.pd

Hokstad, P., Utne, I.B., & Vatn, J. (2012). Risk and vulnerability analysis of critical infrastructures. In P. Hokstad, I.B. Utne, & J. Vatn (Eds.), *Risk and interdependencies in critical infrastructures: A guideline for analysis* (pp. 23–34). Springer.

Humayed, A., Lin, J., Li, F., & Luo, B. (2017). Cyber-physical systems security – A survey. *IEEE Internet of Things Journal, 4* (6), 1802–1831. doi:10.1109/JIOT.2017.2703172

Jaatun, M.G., Albrechtsen, E., Line, M.B., Tøndel, I.A., & Longva, O.H. (2009). A framework for incident response management in the petroleum industry. *International Journal of Critical Infrastructure Protection, 2,* 26–37. doi:10.1016/j.ijcip.2009.02.004

Johnsen, S.O. (2012). Resilience at interfaces. *Information Management & Computer Security, 20*(2), 71–87. doi:10.1108/09685221211235607

Johnsen, S.O., Skramstad, T., & Hagen, J. (2009). Enhancing the safety, security and resilience of ICT and SCADA systems using action research. In C. Palmer & S. Shenoi (Eds.), *Critical infrastructure protection III* (pp. 113–123). Springer. doi:10.1007/978-3-642-04798-5_8

Kamp, G. (2016). *Security convergence in a critical infrastructure – Framework and enablers for successful implementation* [Master's thesis]. Leiden University.

König, P.D. (2021). Citizen centered data governance in the smart city: From ethics to accountability. *Sustainable Cities and Society, 75,* 103308. doi:10.1016/j.scs.2021.103308

Kriaa, S., Pietre-Cambacedes, L., Bouissou, M., & Halgand, Y. (2015). A survey of approaches combining safety and security for industrial control systems. *Reliability Engineering & System Safety, 139,* 156–178. doi:10.1016/j.ress.2015.02.008

La Porte, T.R. (2020). Doing safety … and then security: Mixing operational challenges – Preparing to be surprised. In C. Bieder & K. Pettersen Gould (Eds), *The coupling of safety and security* (pp. 75–85). Springer.

Lee, E.A. (2006, October 16–17). *Cyber-physical systems – Are computing foundations adequate?*NSF Workshop on Cyber-Physical Systems: Research Motivation, Techniques and Roadmap, Austin, TX, USA. https://ptolemy.berkeley.edu/publications/papers/06/ CPSPositionPaper/

Leith, H.M., & Piper, J.W. (2013). Identification and application of security measures for petrochemical industrial control systems. *Journal of Loss Prevention in the Process Industries, 26*(6), 982–993. doi:10.1016/j.jlp.2013.10.009

Lun, Y.Z., D'Innocenzo, A., Smarra, F., Malavolta, I., & Di Benedetto, M.D. (2019). State of the art of cyber-physical systems security: An automatic control perspective. *Journal of Systems and Software, 149,* 174–216. doi:10.1016/j.jss.2018.12.006

Lyngaas, S. (2024, March 19). *Cyberattacks are hitting water systems throughout U.S., Biden officials warn governors.* CNN. https://edition.cnn.com/2024/03/19/politics/cybera ttacks-water-systems-us/index.html

Malmedal, B., & Skotnes, R.Ø. (2019). *Sikkerhetskultur i vann- og kraftsektorene* (NorSIS-rapport). Norsk senter for informasjonssikring.

Mondschein, J., Clarck-Ginsberg, A., & Kuehn, A. (2021). Smart cities as large technological systems: Overcoming organizational challenges in smart cities through collective action. *Sustainable Cities and Society, 67,* 102730. doi:10.1016/j.scs.2021.102730

Norwegian Ministeries. (2019). *National cyber security strategy for Norway.* https://www. regjeringen.no/contentassets/c57a0733652f47688294934ffd93fc53/national-cyber-se curity-strategy-for-norway.pdf

Olenick, D. (2020, April 28). *Israel's water companies suffer cyber-attack.* SC Media. https:// www.scmagazineuk.com/israels-water-companies-suffer-cyber-attack/article/1681580

Patel, S.C., & Sanyal, P. (2008). Securing SCADA systems. *Information Management and Computer Security,* 16(4), 398–414. doi:10.1108/09685220810988 04

Perelman, B. (2018, February 21). The rise of ICS malware: How industrial security threats are becoming more surgical. *Security Week.* https://www.securityweek.com/rise-ics-ma lware-how-industrial-security-threats-are-becoming-more-surgical/

Rasmussen, J. (1997). Risk management in a dynamic society: A modelling problem. *Safety Science,* 27, 183–213. doi:10.1016/S0925-7535(97)00052-0

Renn, O., Klinke, A., & Schweizer, P.J. (2018). Risk governance: Application to urban challenges. *International Journal of Disaster Risk Science,* 9(4), 434–444. doi:10.1007/ s13753-018-0196-3

Skotnes, R.Ø. (2015). Management commitment and awareness creation – ICT safety and security in electric power supply network companies. *Information & Computer Security,* 23(3), 302–316. doi:10.1108/ICS-02-2014-0017

Soomro, Z.A., Shah, M.H., & Ahmed, J. (2016). Information security management needs a more holistic approach: A literature review. *International Journal of Information Management,* 36, 215–225. doi:10.1016/j.ijinfomgt.2015.11.009

Stavanger Municipality. (2016). *Veikart for Smartbyen Stavanger – Visjon, mål og satsing- sområder.* https://issuu.com/stavanger.kommune/docs/veikart_for_smartbyen_stava nger_121/1?ff&e=4979314/53937755

Stouffer, K., Falco, J., & Scarfone, K. (2011). *Guide to industrial control systems (ICS) security, distributed control systems (DCS), and other control system configurations such as programmable logic controllers (PLC)* [Special Publication (NIST SP)-800–882]. National Institute of Standards and Technology. doi:10.6028/NIST.SP.800-82r1

United Nations. (2015). *Transforming our world: The 2030 agenda for sustainable develop- ment.* https://sdgs.un.org/2030agenda

U.S. Department of Homeland Security. (2015). *The future of smart cities: Cyber-physical infrastructure risk.* National Protection and Programs Directorate, Office of Cyber and Infrastructure Analysis. https://www.cisa.gov/sites/default/files/documents/OCIA% 20-%20The%20Future%20of%20Smart%20Cities%20-%20Cyber-Physical%20Infrastruc ture%20Risk.pdf

Wang, L., Törngren, M., & Onori, M. (2015). Current status and advancement of cyber- physical systems in manufacturing. *Journal of Manufacturing Systems,* 73, 517–527. doi:10.1016/j.jmsy.2015.04.008

Wolf, M., & Serpanos, D. (2018). Safety and security in cyber-physical systems and Inter- net of Things systems. *Proceedings of the IEEE,* 106(1), 9–20. doi:10.1109/ JPROC.2017.2781198

Ylönen, M., Tugnoli, A., Oliva, G., Heikkila, J., Nissilä, M., Iaiani, M., Cozzani, V., Setola, R., Assenza, G., van der Beek, D., Steijn, W., Gotcheva, N., & Del Prete, E. (2022). Integrated management of safety and security in Seveso sites – Sociotechnical perspec- tives. *Safety Science,* 151, 1–14. doi:10.1016/j.ssci.2022.105741

9 Streetwise in the artistic city
Jazz, Beats, *Hugs and Bugs*

Kristoffer Berre Alberts and Petter Frost Fadnes

Wise up!

Allen Ginsberg's *Howl* (1956) was one of the first pieces of poetry to provide modern, post-Bloomsbury-bohemian counterculture with a voice. The poem, drawn from hours of field recordings capturing the sounds of New York City, perpetuates a type of urban knowledge the Beats (i.e., Beat poets or Beat movement) referred to as *streetwise*, an emerging, urban ideology based on survivalism (Anderson, 1990), rebellion (Quinn, 2004), non-commerciality, spontaneity (Belgrad, 1998), and how to cunningly manoeuvre the drug-infused madness of a city (Davis, 1990; Pepper & Pepper, 1994). Richard Quinn (2004) described a new generation of urban thinkers and intellectuals (*anti-consumerist-suburbians*) who highlighted 'the active process of improvisation' as a common methodological approach, critiquing, in his words, 'hegemonic post-war passivity while simultaneously fighting its effect' (p. 154). The movement was anti-hegemonic, against all forms of backwardness, set up to mirror real life through the intersubjective and transdisciplinary exchanges between post-war city-dwellers. This echoes through Kerouac's *On the Road* (1957), with jazz protagonists like Charlie Parker on an 'urban pedestal', playing amongst 'the children of the American bop night' (Kerouac, 1957, p. 139). Through Kerouac, we clearly hear (see, feel, and smell) the *sound* of the artistic city as it evolves through street-bound ontologies. This notion of streetwise is connected not just to the Beats but to the rise of youth culture, including the post-war generations' appropriation of urbanity as contemporary, technological, experimental, and cutting-edge.

Smart streets

How is this relevant to a smart city discourse? Although smart cities initially focused on technological innovation solving urban challenges, the term has increasingly gained social and humanistic perspectives of urban citizenship (see Aunemo et al., Introduction). This includes the arts as a 'smart' contributor towards everything from regeneration and sustainability to citizen involvement (see Fisker et al., chapter 6), integration, health, and well-being. Shipman's (2019) chapter 'Smart Art for Smart Cities' highlights art as humanizing to the otherwise

DOI: 10.4324/9781003498650-12

'cold' smart city environment, asking what makes these cities 'liveable' (p. 251). And, when Mostafa Hatem (2023) looked at the 'connection between art and smart', she concluded how the artistic perspective helps us 'feel something'; not to be 'treated as separate' to smart city innovation and science (pp. 301–302). Representatives of the arts are also increasingly contributing to research, innovation, and policy development, like Horizon Europe (EU) and EU policy (e.g., The New European Bauhaus) and the United Nations Sustainable Development Goals, as well as the transformation of STEM (science, technology, engineering, and mathematics) into STEAM (A = arts). Such arts-inclusive policy trends have resulted in various combinations of *smart* and *art*, albeit where few have implemented art into an official smart city policy. A notable exception was the Norwegian city of Stavanger (see Lindland and Matre, chapter 15), which until recently included *urban art/smart art* as the fifth pillar of its thematic focus area (Smartbyavdelingen, 2021; Stavanger Municipality, 2016). Although *Stavanger Smart City* was disbanded at the end of 2023 (see Undheim and Sageidet, chapter 10), it meant that the dedicated department had an yearly annual budget for developing artistic work connected to the other four pillars they were meant to cater to: (a) health and welfare; (b) education and knowledge; (c) energy, climate, and environment; and (d) governance and democracy. Inspired by the Stavanger case, we use this chapter to speculate on the wider potential to 'artify' smart city policies and then test how smart art connections stand up to a streetwise ethos.

Viewed from an artistic perspective, there is of course the danger of stating the obvious: that artistic processes are inherently creative in ways most regard as smart (i.e., in no need of smart city assistance). This is not least apparent within facets of urban counterculture, ranging from 40s bebop, 50s Beats, 60s free jazz, 70s rock, 80s punk, 90s electronica, and 2000s P2P sharing to the various transnational networks and online cultural waves of the subsequent decades. In our context, the challenge is not about creative innovation as much as it is about connecting environments otherwise separated by cultural standings, economics, politics, and power structures. Include artistic contributions towards change and we start seeing the arts' smart city potential. And, although the cityscape-cum-virtual environment we use as a case study in this chapter does not represent 'danger' per se, the analogy of streetwise survivalism still seems apt in the context of countercultures and Beat-like vernacular speech. Streetwise is not just 'food and shelter' (i.e., making a living as an artist) but also the urban expression, experimentalism, and energy Ginsberg (1956) reflects through *Howl*. In a contemporary, post-digital world (Cramer, 2015), we may also add new technologies (Kittler, 1997), the internet (Dubber, 2015), analogue revivals,, and evolving sub-cultures (Fadnes, 2020) as solutions to how current disasters – not least pandemics and wars – initiate creative expressions and streetwise opportunities. The artistic city has already expanded from metropolis to cyberspace, and adapting *beatitude* to a sphere between the analogue, digital, metropolitan, and rural, the following text is about creative problem solving and streetwise adaptation in a smart city context. We base our argumentation on an artistic case study; more specifically, a piece of music which shifts between a Stavanger origin and online collaboration from multiple locations. Recorded during April 2020, the music was developed as a desperate response to pandemic lockdowns and a music scene in crisis. Titled *Hugs and Bugs* (*HaB*), it is a collage-type musical composition patched together by blocks of music, contributed by musicians in isolation across Europe. Initiated and curated by musician/improviser Kristoffer Berre Alberts, the 14-member group was aptly named Block Ensemble. Invited based on their knack for artistic problem solving, the musicians completed their recordings without technical assistance or specific compositional guidelines, and only post recording did Alberts get technically involved, using a digital audio workstation to edit the material into a sonic collage. The album *Hugs and Bugs* (Block Ensemble, 2021; Figure 9.1) was released digitally and physically in 2021.

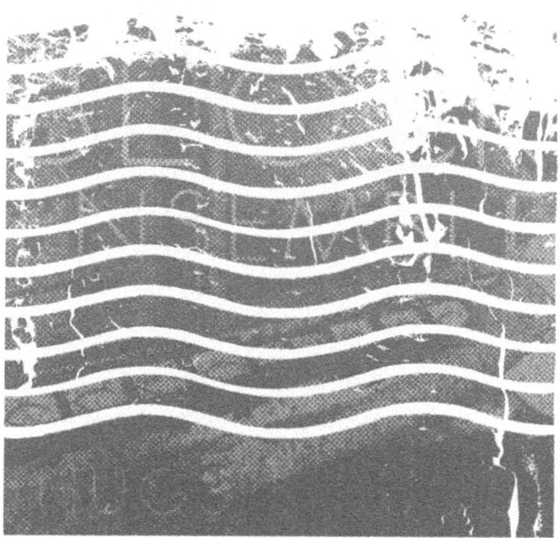

Figure 9.1 CD cover of *Hugs and Bugs* by Block Ensemble
Source: Cover design by Kjetil Brandsdal, courtesy of Clean Feed Records.

A ten-point manifesto

African and Black Diaspora scholar Amor Kohli (2016) makes an excellent reflection on the artist Ted Joans' initiation of the phrase 'Bird Lives!' – an homage to the then recently deceased Charlie Parker (1920–1955). This graffiti – a.k.a. what we now call street art – was seminal in coining the phrase 'Bird Lives!' as a symbolic gesture towards civil rights, urbanity, bebop, and the jazz legacy. 'Bird Lives!', Kohli argued, 'emphasizes Bird's music as a generative force, transforming how we now "sound"' (Kohli, 2016, p. 99). With an emphasis on *sound*, we see – or rather hear – how streetwise is not just hipster, jive, and trendiness but a more substantial formation/affirmation of identity within urban countercultures. Alexandre Ferrere (2020) referred to the Situationists, specifically Guy Debord's *psychogeography*, and how the 'psychogeographical poems of Ginsberg are the embodiment of an inner struggle', a way of 'transcribing the world' (p. 14).

Psychogeography is both inner (psyche) and outer (geography), marred with the poetic landscape of the artist. If smart art – as either policy, symbolism, or a form of creative tool – can capture the *sound* of urbanity, it gains momentum, meaning, and purpose beyond empty policy jargon. To 'set the record straight', we therefore propose a list of ten Manifesto Points (MPs 1–10) of what smart *is* and *isn't* in an artistic context (equal pros and cons) before we contextualize the statements in the presence of *HaB*.

MP1 The semantic meaning of *smart* (intelligent, neat, trendy, computer-controlled, quick; Oxford Dictionary, Stevenson, 2010) has no meaning within the artistic field and should not be used in an artistic context.

MP2 Incorporating smart art as part of a smart city policy emphasizes artistic practices and expression as a vital part of a modern, innovative city environment.

MP3 Smart implies the presence of dumb and questions *who* or *what systems* should have the power to make that distinction (smart city CEOs?).

MP4 A smart city ethos is about user-friendliness, accessibility, clever forms of mediation, and the connection to local, urban cultures, and the artistic communities can learn from this – and force them out of their secluded 'bubbles'.

MP5 When it comes down to it, smart art is *only* about the potential for commercial value, specifically in which way the artistic fields are able to connect to ground-breaking industries (e.g., technological innovation).

MP6 Within the fields of experimental art, we have no idea which expressions remain secluded experiments and what history deems to be important contributions to the evolution of humanity (and the arts) – ergo, we have no idea whether they are smart or dumb.

MP7 An emphasis on smart art will help integrate artistic dimensions into *all* aspects of our lives and put greater demands on commercial industries (design, city development, innovation) to also include artists in their development and decision-making processes.

MP8 Smart art empowers art and shows that all art is inherently smart (creative, innovative, mediated, connected, etc.).

MP9 Smart art within a smart city context is 'artwashing', exploiting artistic alibies for a cynical gain (Ginsberg would *not* approve).

MP10 The **SMART** logo – including *sm(art)* – is a clever branding exercise for the development of urban cultures, and the arts are forced to 'ride along' rather than being left behind.

Although the list is ambiguous, it certainly takes the arts discourse seriously and engages in the instability of smart(ness) – not just related to the arts – but the smart city discourse in general.

Becoming smart(?)

Through rereading/recontextualizing urbanity, *HaB* moves away from city streets to embrace the other-worldly surroundings of digital connections and lonesome physical disconnections. Juxtaposed with MPs 1 to 10, such a shift generates valid concerns: does the idea of smart make sense in an artistic context (MPs 1, 3)? Is it smart to begin with (MP8) or does it have the potential to *become* smart (MP4)? Can we foresee whether art is perceived as smart(er) after our time (MP6)? Should the artistic city strive to embrace smartness more enthusiastically (MPs 2, 7), or is the whole thing just a policy-based marketing ploy (MPs 5, 9, 10)? Historically, we know that some of the great defining artistic expressions are represented or articulated through crisis, segregation, and war. Does it follow that brutal internal, political, and violent struggle is part of the smart art discourse? Perhaps the answers lurk somewhere in the asking...

Sharing cultures

The meaning of smart in a performative context tends to boil down to elements of decision making: on the one hand, making the 'right' decisions (i.e., not dumb), on the other, *trying* to make the 'right' decisions. And, as we know, trying is both much more fun and the only non-essentialist perspective that makes sense in an artistic context. It is often, in other words, about experimentation (see Lorentzen and Langhelle, chapter 4), which, fun and playful as that is, invariably includes terrible mistakes and utter failures. By imagining a trying/failing axis based on artistic *doings*, we start understanding a process based on the cognitive nature of learning (Pressing, 1984), knowledge production (Fadnes, 2021), inside knowledge (Ingold, 2013), and shared knowledge (Richardson, 2016) – all through a long line of temporary events, misunderstanding, failures, and intermediaries. The reflection of all this can be summarized as a *point of reference*, relating to identity formation/affirmation, how the world sees you, how you want to be seen, and, in a Butlerian sense (Fadnes, 2020), *what you do.*

Although it is hard to draw a coherent line between a point (or points) of reference as obvious origins, we nevertheless sense, to borrow from ethnomusicologist Ellen Koskoff (1982), crucial relationships between 'cultural and individual meanings' (p. 353) as the foundation of a complex network of artistic knowledge. To help untangle this, musicologist Bruno Nettl (2015) referred to *musical models* (styles, schools, or genres), highlighting connections between knowledge and references as their precursors. In our context, Nettl shows how the improvisational points of reference behind the models are connected in meaningful ways, establishing forms of embodied knowledge. In South Indian classical music, for example, the strict rhythmical and melodic references represent a stringent network outlining the improvisational foundation of the music. These references-cum-models are often groomed from childhood through sociocultural and psychogeographic connections: family members, schooling, communities, and place. By seeing musical improvisation as a skill originating from specific reference points, we may therefore also start experimenting with and manipulating these skills in unusual contexts. Hence, we move into a territory where all the MPs play with the potentials and opportunities of art: urban-cultural development (MPs 2, 4), artistic integration and mediation (MPs 7, 8), commercialization and jobs (MP10).

In addition to the sum of the references which artists amass through a lifetime, the artists are driven by an urge to develop new meanings, impact change, be relevant, and achieve a sense of self-fulfilment and personal development through their work. In addition to Beat poetry and the 40s bebop, facets of 50s and 60s jazz are examples of self-centrism ('art for art's sake'), marred with symbolism of a utopian world or, at the very least, a more *just* world against a segregated United States (Lewis, 2008). In addition to serving as a representative of political change and artistic innovation in *the West*, jazz represented a counterculture of jazz enthusiasts and political radicals behind the iron curtain in *the East* (Fadnes, 2018; Hakobian, 1998). To some, American jazz symbolized the enemy; to others, emancipation and democracy. The same sound and genre but different references to whatever sociocultural contexts make up a jazz-performative identity. Further east, in the late 70s and 80s, Japanese musicians who identified as punk rockers engaged in various kinds of anti-music (Novak, 2013), presumably borrowing from dadaism, fluxus, situationism, and the like. Their recorded outputs were shipped off to North America, re-released, and subsequently labelled *noise* music, a term that shows the music falling outside of the etymology meaning to play quiet or in harmony. Conversely, it created an internal gap between the factions of Japanese musicians; some identified with the North American classification of (their) *noise*, whilst others found it alienated from the origins of their artistic intent. Nevertheless, the music they produced – and keep producing – is still globally categorized as *noise*, or *japanoise*, with further sub-genres such as *harsh noise, noise rock, ambient noise, pink noise, white noise*, and others.

In considering both analogue and digital sharing platforms, we recognize the vastness of the tools we have at our disposal to experience music globally and beyond borders and, as technologies become ever cheaper and more accessible, how these tools get democratized and accessible beyond social hierarchies. In fact, as the world embraced innovative technologies, streetwise had to remap its psychogeographic connections – even making us question the logic of upholding place-bound meanings within a global, digital, sharing-based playing field (MPs 4–7). The introduction of audio cassettes in the 70s, for example, made it easier for remote artists who were without label-backing. These artists were helped by, in Novak's words, 'new social and economic relationships around sound recordings, allowing individual users to reproduce, remix and distribute their own material' (Novak, 2013). Chain Mail Collab (Das et al. 1988), for example, created and composed what they saw as an improvisation-based *circle of cooperation*: originating in a single musician's recorded material being mailed on audio cassette to the next person, who would then add to the tape, a process potentially repeated in perpetuity. This media-circulative method of creating music was prominent within the so-called noise music generation, and the sonic content was usually electronic signals modified and transformed through guitar amplifiers, synthesizers, and other electronic supplies that change the nature of the sounding source. In addition to the electronic manipulation process, this represented a new improvisational direction compared to the site-specific, real-time, interactive nature of jazz. The sharing of audio cassettes as an improvisational practice differed from the ensemble ideology of the jazz canon, questioning whether improvisation was now lifted *beyond* the sound of the artistic city altogether.

With reference to Nettl's (2015) models, the use of the term *noise* comments on how we understand urban-artistic ecosystems and how they can expand into something undefined, constantly in flux and 'unplaceable'. This also examines how we recognize and experience improvisation beyond momentary/spontaneous/real-time points of reference (i.e., gestures, incidents, events). When a live improvisational dialogue works at its best, we often refer to *flow* (Csikszentmihalyi, 1990), improvisational events developing/flowing through time unbroken. *HaB* breaks the ideal of a flow-based timeline and rather reconstitutes – in an adaptable manner – improvisation in an age of digital sharing cultures. The piece challenges established knowledge, how to utilize material from isolated sources, and how to virtually cross borders beyond established models. Smart? Not smart? Would Ginsberg approve?

Hugs and Bugs

The global pandemic that hit in 2020 forced us into isolation and made us re-evaluate how we work as musicians. Deprived of otherwise common referential points related to collectivity and liveness, 'being on our own' put knowledge, competence, flexibility, and even identity to the test. Although some embraced the situation (peace and quiet, free of everyday distractions, time to think and acquire new skills), for many, it was unexpected and challenging. *HaB* was one of many artistic outcomes around the world sparked by lockdown, and although it 'stands on the shoulders' of 70s and 80s cassette cultures, COVID limitations demanded new, innovative solutions.

The circle of cooperation making up *HaB* consists of 14 musicians contributing individually from wherever they were stuck in pandemic isolation. Although the process is inspired by noise/Japanoise work methodologies, almost all the members consist of performers from outside the noise genre, leaning more towards forms of jazz, acoustic improvised music, and composed contemporary music. The group was curated by Kristoffer Berre Alberts and not, with regards to 'noise authenticity', as a curatorial chain. Emails were instead sent out asking for solo improvisational recordings, which were then posted to Alberts as digital sound files containing diverse sound qualities. The material was then (re)composed, spliced, and edited, turning the isolated acoustic material into new wholes (Figure 9.2). Breaking with established noise models, the musicians pursued the acoustic qualities of their instruments, setting up a sense of false, 'synthetic', or emulated interplay.

As a start, an email was sent out to the musicians, asking them to follow a series of six tasks:

1   Three-note melody.
2   Short and loud, slow and low.
3   Massive and fast.
4   Space is the place and far away.
5   Too many hugs or too many bugs.
6   Dubbing/audio guide.

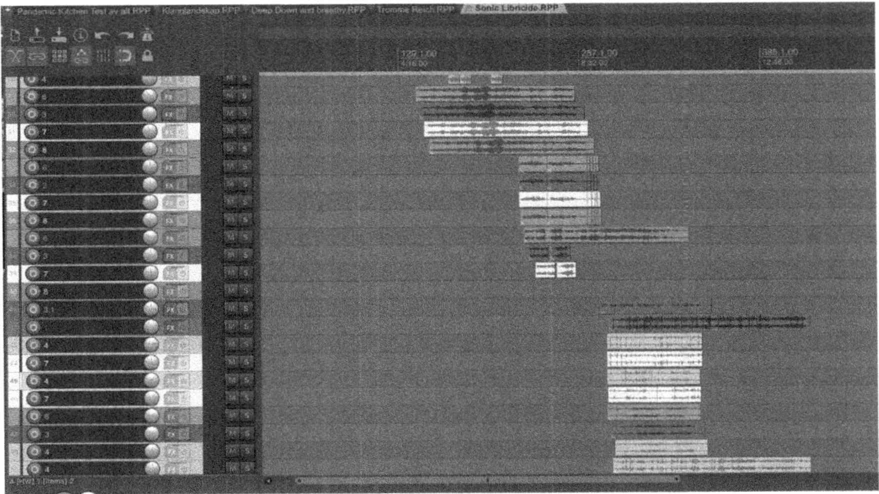

Figure 9.2 Finished edit of the A side of *HaB* LP. The recorded material was cut and blocked in the digital audio workstation software Reaper.

Play with one of the above on your headphones or by memory. If you like, you can do several tracks and feel free to use your own unique dubbing technique. You can also play along with your favourite record.

The motivation for playing live is for most improvisers the social meeting points between co-players and an audience. *HaB* initially robbed its participants of this; although Alberts gave the musicians composed tasks to work from, including the possibility to completely disregard the task, be punk; that is, *give it the finger*. In speaking to musicians afterwards, it seemed COVID-19 had motivational impetus; the invitation to join a community – albeit virtual – was motivation enough. Based on an aesthetically common ground between the musicians, the sum of previously amassed experience and knowledge from live, improvised music made the record sound coherent beyond layers of blocked sounds. *Downbeat Magazine*'s Martin Longley seems to recognize this:

> The results are gripping and calming, in turn, involving some of the most impressive improvising heard in recent times, even though these elements are 'comprovisations', filtered by the mind of Alberts.
>
> (Longley, 2022)

Longley's use of *comprovisation* normally relates to a process where improvisation and composition meet outside canonized genres, and although the piece is certainly pushing genre boundaries, *HaB* gains a sense of identity through being curated and composed, organized, and filtered through the mind of one single person. And despite it sounding very much like a physically present, in-sync collective, in most cases the sound files were nothing more than a chaos of different ideas, far removed from the ideals of streetwise jazz clubs. In fact, if we see streetwise as knowledgeable insights into the vast array of city-living reference points (with communal connections), it is hard to place *HaB* inside the artistic city. Brutal and unromantic as it sounds, it seemed unimportant that Alberts was situated in Stavanger – not least due to the pandemic lockdown. Removed from the streets and the delicate wisdom connected to them, it was imperative for the project to find its ground zero, its *tabula rasa*, or, at the very least, a common ground for building the artistic process – without compromising artistic diversity or breaking ethical boundaries. The curatorial process of putting together an ensemble with a feel for digital manipulation of acoustic material was essential. They all had to sign up for a process by which their precious recordings would be disordered and recontextualized. This was the danger, the open poetry, the agreed *street etiquette*.

Devastating as it was, the enforced lockdown became the point of departure (MP 6): musicians agreed to the project because they had time and wanted to stay active with their playing, instinctively feeling the need for experimentation through difficult times. Otherwise used to collectivity as the foundation of their daily practice, they were now left by themselves, forced to find new solutions to their unplanned predicament. Fortunately, they had the infrastructure (instruments, microphone, computer, sound card) and the knowledge to keep producing a musical output. The contexts varied, however, and the idiosyncrasies of their surroundings mirror the construction of the piece and how it forced a rethink/rework of points of reference otherwise drawn from real-time, live concerts. The creative advantage was that time and space could be manipulated, deconstructed, and rebuilt; streetwise knowledge making its presence as a 'living, breathing'

Figure 9.3 Ingebrigt Håker Flaten recording for *HaB* at his mother's house in Oppdal, Norway.
Source: Grainy screenshot from video recording by Flaten, June 2020.

testament to creative problem solving. Most important, the piece became a document of how different improvisational ideas, diversity, and chaos merged into a collective experience, offering the musicians a new way to process improvisation, genre, and relational frameworks.

When live concerts recommenced after a long period of lockdown, many musicians, rather surprisingly, seemed to have enhanced their improvisational skillset. Desperate challenges help break patterns, enforce new thinking, and even advance new knowledge. (Is desperate art still smart? [MP 8]). It seemed that *HaB* and other art projects (see, e.g., Thorley, 2023) cultivated during lockdown actively invited musicians into a new sustainable, digital street etiquette, in this case fuelled by circulative and reusable improvisatory knowledge.

Stavanger streetwise or Stavanger 'streetwise'

Being a born-and-bred Stavangerian entitles you to the slang-based label *siddis*. Alluding to the traditionally anglified nature of the city, *siddis* is a local variant of the English word *citizen*. In fact, the term, as referred to in 1920s sources, was *Stavanger-Citissen* – a citizen of Stavanger. In the Porto Santo Charter (2021) on cultural citizenship, democracy (or democratization) is highlighted as the highest virtue, its goal to engage citizenship across Europe. 'Democracy is a dynamic social methodology', the report states, providing 'a voice and a choice', which, through respect and diversity, 'relies on the cooperative intelligence of the community' (Porto Santo Charter, 2021, p. 4). *HaB* represents a streetwise solution to an age-old problem: how to survive in the face of adversity. The democratization process, as highlighted by the Porto Santo Charter, is in our case ensured by musicians utilizing the means at their disposal. By all accounts, this is an activist stand, a deep-rooted need, stretching back to Ginsberg's call to arms, in which *HaB*, to reborrow from Ferrere (2020), is a *psychogeographical poem* of its day: 'transcribing the world [...] into a new vocabulary or context' (p. 14). *HaB* is thereby a sonic-digital imitation of 'Bird Lives!' inscribed by individuals in solitude but meticulously collated and curated as a virtual community by a siddis,[1] a citizen of Stavanger.

Digitalization in music production is here to stay, and so are its users. Put otherwise, most of us live and breathe as citizens in a place we call home, and that place – city, village, recording studio, online community, or jazz club – has as much impact on *our doings* as the technology we utilize. In *HaB*, Alberts drew on musicians from diverse places, urban and rural, domestic and international: from Oppegård and Oppdal, to Stavanger, Trondheim, and Oslo, and further afield to Berlin, Amsterdam, and Stockholm. Although physical contact became immaterial, the piece was nevertheless founded on a shared, collective, and ultimately chaotic (Borgo, 2007) common ground. The entanglement of commonality (city, space, or place) becomes the sum of all points of reference (departed *from* or referred *to*). As keen listeners in an audience, or in front of our living room speakers, we hear with limited comprehension but tend to understand and experience what references are at play within the music. By presenting a case study linked to a specific artistic practice, and to situate that in a smart city discourse entangled by theories, manifestos, hypotheses, and parallel artistic events, we problematize the reader's subjective idea of how they value the activities of the artistic city. In the spirit of MP6 – *not knowing whether art is smart or dumb* – we cautiously consider *HaB* to be a smart art project but have no idea whether it will stand out as a secluded experiment or whether it will have artistic impact and value on its own merits. In fact, it was essential that this project evolved from musical and artistic practice – not to force existing theory on practice but to undertake a type of *bottom-up* theory development. Rising from practice into the streets (or the digital-collaborative world), we looked for whatever wisdom, mistakes, wrong turns, and subtle approaches the project demanded. If this non-theoretical approach is considered smart, and by referring to MP8 – *that all art is inherently smart* – we may iron out the frictions: from smart- or non-smart art to just simply art.

Having to deal with unstable variables such as fleeting audiences, socio-political fluctuations, pandemics, wars, and many other unknowns brings us back to our manifesto and how the definition of smart in the artistic city is distracting at best, at worst, outright stupid. Streetwise, Ginsberg, 'Bird Lives!', and pandemic restrictions – are we any wiser? Perhaps *HaB* reveals smartness as the reconstitution of sharing cultures, which, in a post-digital spirit, utilizes what is already here in a sustainable, effective, and creative way. Within the Bourdean *open space of interest and investments* that art represents, smart (in a streetwise sense) is *already* outdated, past its 'sell-by date', useless for future action, beyond already established models and references. Realizing the essence of smart art is, therefore, to look at the underlying factors of smart processes, smart intermediaries, and smart outcomes, including how the art is perceived and, maybe most important, how it is understood and expressed by artists themselves. We can investigate flowing timelines driven by feedback loops, previously acquired knowledge, noisy methodologies, a trying/failing axis, and instrumental objectives that are forced to adapt to processual change. But to even attempt to label something as smart art in a manner befitting the streetwise clan of Beats, we must account for perspectives of time and place and realize our limitations.

Acknowledgement

We thank all the musicians participating in the *Hugs and Bugs* project for their excellent musical contributions and the Smart City Research Network at the University of Stavanger for their necessary support throughout the research period.

Note

1 Alberts only moved from Oslo to Stavanger a few years ago, but in the spirit of globalization, migration, and free movement, this is a small detail we choose to ignore.

References

Anderson, E. (1990). *Streetwise: Race, class, and change in an urban community*. University of Chicago Press.

Belgrad, D. (1998). *The culture of spontaneity: Improvisation and the arts in postwar America*. University of Chicago Press.

Block Ensemble. (2021). *Hugs and bugs* [Album]. Clean Feed Records.

Borgo, D. (2007). *Sync or swarm: Improvising music in a complex age*. Continuum.

Cramer, F. (2015). What is 'post-digital'? In D.M. Berry & M. Dieter (Eds.), *Postdigital aesthetics* (pp. 12–26). Palgrave Macmillan.

Csikszentmihalyi, M. (1990). *Flow: The psychology of optimal experience*. Harper Perennial.

Das, Jerman, J., Hudak, J., & PBK. (1988). *Chain Mail Collab June 28, 1988* [Album]. https://pbksound.bandcamp.com/album/chain-mail-collab

Davis, M. (1990). *Miles – The autobiography*. Picador.

Dubber, A. (2015). Collective practice and digital mediation. In N. Gebhardt & T. Whyton (Eds.), *The cultural politics of jazz collectives: This is our music* (pp. 219–235). Routledge.

Fadnes, P.F. (2018). Free spirits: The performativity of free improvisation. In N. Gebhardt, N. Rustin-Paschal, & T. Whyton (Eds.), *The Routledge companion to jazz studies* (pp. 443–453). Routledge.

Fadnes, P.F. (2020). *Jazz on the line: Improvisation in practice*. Routledge.

Fadnes, P.F. (2021). Wordplay: Negotiating the conservatory 'culture clash'. In M. Kahr (Ed.), *Artistic practice as research in jazz and popular music: Positions, theories, methods*. Routledge.

Ferrere, A. (2020). Creative environments: The geo poetics of Allen Ginsberg. *Humanities*, 9(3), 1–15.

Ginsberg, A. (1956). *Howl and other poems*. City Lights Books.

Hakobian, L. (1998). *Music of the Soviet age, 1917–1987*. Melos.

Ingold, T. (2013). *Making: Anthropology, archeology, art and architecture*. Routledge.

Kerouac, J. (1957). *On the road*. The Viking Press.

Kittler, F.A. (1997). *Literature, media, information systems: Essays* (J. Johnston Ed.). Psychology Press.

Kohli, A. (2016). Sounding across the city: Ted Joans's *Bird Lives!* as jazz performance. In D.R. Geis (Ed.), *Beat drama: Playwrights and performances of the 'howl' generation*. Bloomsbury.

Koskoff, E. (1982). The music-network: A model for the organization of music concepts. *Ethnomusicology*, 26(3), 353–370.

Lewis, G.E. (2008). *A power stronger than itself: The AACM and American experimental music*. The University of Chicago Press.

Longley, M. (2022, January). Clean Feed's 20th anniversary. *Downbeat*.

Mostafa Hatem, F.A.F. (2023). The role of art in smart cities research and making. *IET Smart Cities*, 5(4), 291–302.

Nettl, B. (2015). Thoughts on improvisation: A comparative approach. In A. Heble & R. Caines (Eds.), *The improvisation studies reader: Spontaneous acts* (pp. 69–78). Routledge.

Novak, D. (2013). *Japanoise: Music at the edge of circulation*. Duke University Press.

Pepper, A., & Pepper, L. (1994). *Straight life: The story of Art Pepper* (Updated ed.). Da Capo Press.

Porto Santo Charter. (2021). *Culture and the promotion of democracy: Towards a European cultural citizenship*. https://portosantocharter.eu/wp-content/uploads/2021/05/PortoSantoCharter.pdf

Pressing, J. (1984). Cognitive processes in improvisation. In *Advances in psychology* (Vol. 19, pp. 345–363). Elsevier.

Quinn, R. (2004). Jack Kerouac, Charlie Parker, and the poetics of beat improvisation. In J. Skerl (Ed.), *Reconstructing the beats* (pp. 151–167). Palgrave Macmillan.

Richardson, L. (2016). Sharing knowledge: Performing co-production in collaborative artistic work. *Environment and Planning A: Economy and Space*, 48(11), 2256–2271.

Shipman, H. (2019). Smart art for smart cities. In M. Mateev & P. Poutziouris (Eds.), *Creative business and social innovations for a sustainable future*. Springer. https://doi.org/10.1007/978-3-030-01662-3_27

Smartbyavdelingen. (2021). *Halvårsrapport vår 2021*. https://www.stavanger.kommune.no/samfunnsutvikling/smartbyen-stavanger/halvarsrapport/halvarsrapport-var-2021/#17649

Stavanger Municipality. (2016). *Roadmap for the smart city Stavanger – Vision, goals and priority areas*. https://www.stavanger.kommune.no/siteassets/samfunnsutvikling/planer/engelske-planer/roadmap-smart-city-stavanger-2016.pdf

Stevenson, A. (2010). *Oxford dictionary of English*. Oxford University Press. doi:10.1093/acref/9780199571123.013.m_en_gb0783660

Thorley, M. (2023). The pandemic as catalyst for remotivity in music. In M. Agamennone, D. Palma, & G. Sarno (Eds.), *Sounds of the pandemic*. Routledge.

Part III
The lived smart city

10 Implementation of the smart city concept in Stavanger Municipality – from a global idea to local practice

Ståle Undheim and Barbara Maria Sageidet

Introduction

A number of Norwegian cities and municipalities have worked on projects within the *smart city* concept since around 2012 (Agenda Kaupang, 2018; Vedeld, 2022). In 2014, Stavanger Municipality's work with urban development and the *smart city* concept (Albino et al., 2015) achieved EU funding for the Triangulum project, together with Manchester and Eindhoven (Evans et al., 2021; see also Aunemo et al., chapter 1, and Müller, chapter 2). Thus, the implementation of Stavanger as a smart and sustainable city started in 2015, with the creation of a strategy document in the format of a roadmap for the smart city work in the municipality (Stavanger Municipality, 2016). In 2017, along with a systematic implementation, Stavanger became the first city municipality in Norway with its own smart city office (SC office), which hired four new professionals with interdisciplinary backgrounds. Its role was to implement and carry out the measures of the roadmap (Stavanger Municipality, 2016). In Norway, as well as internationally, there is still restricted knowledge about fruitful ways to implement the *smart city* concept and about the possible role of the administrational organization in this regard (Lebrument & de La Robertie, 2020; Meijer & Bolivar, 2016). Based on a master's thesis (Undheim, 2020), this chapter qualitatively explores what happens when an idea meets the field of practice (Christensen et al., 2015; Røvik, 2016), with the research question: What happened with *smart city* ideas under their implementation in the Stavanger Municipality, in view of selected organizational change theory?

This qualitative case study explores how the SC office, established for the smart city implementation in Stavanger, coped with the implementation process in practice, considering respectively rational, natural, new institutional, and translation theory perspectives of organizational change (Christensen et al., 2015; Røvik, 2016), by means of a document analysis and a qualitative questionnaire of key persons. Findings reveal restricted and slowly developing organizational changes, due to multiple challenges.

The smart city concept and organizational change

The *smart city* concept has emerged internationally with various definitions (Albino et al., 2015); however, all include that a *smart city* increases the quality of

DOI: 10.4324/9781003498650-14

life of its citizens (Giffinger et al., 2007; see other chapters in this book; for example, Müller-Eie, chapter 7). In the Nordic countries, organizations and municipalities, including Stavanger (cf. Stavanger Municipality, 2016), have gathered and further developed the concept into a Nordic *smart city* model based on Nordic social values (DOGA, 2019; Kangas & Kvist, 2019). This model acknowledges citizen participation, inclusion, cooperation, and co-creation, as the 'soft' dimensions of the *smart city* concept, equal to the 'hard' technology-related dimensions (Wataya & Shaw, 2019), in line with the Global Sustainable Development Goals (Target 11.3; Norwegian Ministry of Foreign Affairs, 2016), to become more sustainable, productive, and adaptable (Clement et al., 2023; DOGA, 2019; Haarstad, 2016). Embedded in this background, the Stavanger Municipality defines the *smart city* concept at a high level of abstraction (Undheim, 2020).

There is still little knowledge about how smart city governance systems are organized and what they need to succeed in becoming 'smart' (Lebrument & de La Robertie, 2020). Mendoza Moheno et al. (2017) identified organizational challenges faced by smart cities, including organizational structures, spirits, learning, change resistance, and the ability to transform and innovate. Asker Guenduez and Mergel (2022) underlined the importance of capabilities like evaluation, innovation, and integration, as well as readiness of the city administration to drive smart city transformation. According to Clement et al. (2023), many local administrations see the *smart city* concept as an appropriate paradigm for advancing sustainable infrastructure changes at the local level.

Theories of organizational change

The rational perspective

The rational perspective, grounded in Weber's (1922/1968) concept of systematic calculation to maximize efficiency, has been put under pressure for change in recent decades (Røvik, 2016). It views organizations as tools to achieve goals as effectively as possible (Scott & Davis, 2000; Zey, 2015). Formal structures, rules, and norms play important roles in this perspective, as they shape and guide what individuals undertake and how they focus on solutions and challenges, independent of who holds the positions in the organization (Zey, 2015). A leader, if appropriate, will have the leeway to change the organization in the direction of what best serves the purpose (Christensen et al., 2015). Changing ideas can be quickly connected to and implemented in the organization using ready-made 'tools'. However, if the new ideas are threatened, they can be harmonized with established values and practices of the organization (Røvik, 2016).

The natural perspective

The natural perspective, which emerged in the 1970s, sees organizations not as purposeful instruments for efficient and almost mechanical production but rather as man-made units that strive to adapt and survive in their particular circumstances (Gouldner, 1959; Scott & Davis, 2000). Culture, including values, informal norms, and myths, plays a central role in this perspective, while rationality is muted. Decisions are made based on experience and assessment of what is culturally appropriate (Christensen et al., 2015) and might be top-down or bottom-up (Easterly, 2008).

Another central aspect of the natural perspective is the concept of *institutionalization*, the process by which an organization adopts special characters and achieves distinctive competences, or maybe develops an incapacity (Scott & Davis, 2000; Selznick, 1996). A formal organization will develop informal norms and values in addition to the formal ones, and these may be sources of strength or weakness (Scott & Davis, 2000; Selznick, 1996). The organization then becomes more complex and less flexible. It also may become more resistant to new changes and demands through building up experience-based knowledge, structures, and capacity (Scott & Davis, 2000).

The new institutional perspective

The new institutional perspective, which emerged in the 1980s, focuses on the organization's outer frames and surroundings and sees organizations as part of, and influenced by, their environment (Powell & DiMaggio, 1991; Røvik, 2016). This perspective addresses both the consideration of efficiency, which often is established as formal and experience-based solutions, and the organization's ability to incorporate ideas and recipes that are perceived as modern at any given time or assumed to give the organization external legitimacy to gain acceptance from the environment, without allowing these ideas to necessarily change daily operations (Meyer & Rowan, 1977). The leaders thus receive ideas but they do not put them into use; the result is then decoupling. This is assumed to happen as part of the handling of all the complex and intense tasks attributed to leaders today, serving as a survival strategy (Powell & DiMaggio, 1991; Røvik, 2016). In situations with chaotic, problematic, or competing preferences, solutions may be linked to problems merely by chance, out of the mix – or 'garbage can' – of existing problems and solutions (Cohen et al., 1972).

Translation theory

Translation theory is based on the introduction of the 'sociology of translation' to organizational theory (Czarniawska & Joerges, 1996; Latour, 1987). Usually, knowledge transfer from a source to target units aims to recreate observed and desired practices and/or results in the context of the target unit (Røvik, 2016). However, the power behind the 'travel' of such ideas stems from the richness of interpretations that the idea generates in the various actors within the target unit or network (Latour, 1987). According to translation theory, transfer involves transport, copying, transformation, and 'editing' (Røvik, 2016). For a successful transformation, good translations are necessary, which require translator competence and knowledge about the organization's reform history. The translator must also have knowledge, courage, patience, and strength and collaborate with the formal management (Czarniawska & Joerges, 1996; Røvik, 2016).

Methodology

This chapter qualitatively explores the implementation of the SC office in Stavanger Municipality from spring 2018 until spring 2020, especially regarding its working methods and the degree of implementation of ideas of the *smart city* concept (DOGA, 2019; Stavanger Municipality, 2016) through ongoing projects (Undheim, 2020). This case study investigates links between a current situation and a selected theory (Thomas, 2011). For this purpose, four assumptions were elaborated, linking *smart city* implementation in the Stavanger Municipality respectively to four alternative theoretical perspectives on organizational implementation: the rationale, the natural, the new institutional, and the translation perspective.

This chapter draws on data from a qualitative document analysis of four semi-annual reports of the Stavanger SC office between spring 2018 and autumn 2019 (available at www.stavanger.kommune.no) and from semi-structured in-depth interviews with three key people from this office (Undheim, 2020). The data were analyzed by writing narrative summaries (Chase, 2013) within a hermeneutic approach, considering influences of the researcher's (and first author's) preunderstandings, beliefs, and biases (Aase & Fosså skaret, 2014; Blaikie, 2010; Kvale, 1997) related to his role as a smart city manager in an adjacent municipality during the study (Undheim, 2020).

Through qualitative coding (Aase & Fosså skaret, 2014) of these summaries, the empirical findings were evaluated and categorized with the help of four assumptions, related to the selected organizational change theories (Christensen et al., 2015; Røvik, 2016) and operationalized by a guiding question and two to four hypotheses or checkpoints (see Table 10.1). Further analysis explored the data and evaluated whether they supported the respective hypotheses/checkpoints or not (see Table 10.1) to find out how far the selected organizational implementation theories were reflected in the data material.

Analysis and discussion

Table 10.1 provides a summary of the analysis of the smart city implementation, considering four alternative theories, while the discussion includes excerpts from the interviews. The analysis indicates that none of the examined assumptions could be dismissed from an overarching perspective, neither generally for the SC office nor for the (at that time) ongoing projects included in the analysis (Table 10.1). However, for some of the assumptions the data show stronger explanatory power and links to them.

Table 10.1 Overview showing checkpoints assessed against what has happened and which assumptions seem to have been supported in the study.

Implementation theories	Hypotheses (checkpoints)	Are the hypotheses supported by the data?
Assumption 1: Ideas were connected quickly 'Rational' implementation	1.1: The SC office and smart city projects are seen as tools for efficient goal achievement. 1.2: The decision is made by the municipality's political leadership 1.3: Management has made a detailed plan 1.4: Visible results quickly come into place	1.1. Partially 1.2. Supported 1.3. Partially 1.4. Partially
Assumption 2: Ideas were repulsed 'Natural' implementation	2.1: Culture-based norms and values remain 2.2: The organization has an inherent culture of change 2.3: Tacit knowledge – the projects meet resistance 2.4: Bottom-up perspective, fails	2.1. Partially 2.2. Partially 2.3. Partially 2.4. Partially supported
Assumption 3: Ideas were decoupled 'New institutional' implementation	3.1: *Smart city* is taken into the organization but no further; that is, is decoupled 3.2: The expectations from the institutional environment do not correspond to daily practice 3.3: Just talk, not practice	3.1. Partially 3.2. Partially 3.3. Partially
Assumption 4: Ideas were translated and adjusted Translated implementation	4.1: The skilled translator has experience and authority 4.2: Translation based on rational considerations	4.1. Partially 4.2. Partially

Source: Adapted from Undheim (2020).

Discussion

The decisions from the municipality's top management to establish an SC office and to develop a smart city roadmap (Stavanger Municipality, 2016) can be considered effective tools to implement the *smart city* ideas in accordance with national laws and local goals and to solve complex problems and carry out *smart city* tasks, consistent with a purposive rational perspective (Scott & Davis, 2000; Zey, 2015); see assumption 1 (Table 10.1). The SC office received much attention around the projects they were testing out. The interviews confirmed that the SC office had, together with certain projects, received remarks at both local and national levels that underscored a quick and rational connection (Scott & Davis, 2000) in the initial phase. Such social authorizations for the work with smart city implementation (Røvik, 2016) contributed to smart city Stavanger's growing local, national, and possibly international reputation.

> Smart City Stavanger has received many accolades. I don't think the SC office would have received all these awards if they hadn't been so good at communicating.
>
> (Employee at the SC office)

In the beginning, the smart city projects appeared dominated by the technical or 'hard' dimension (cf. Wataya & Shaw, 2019), also including networking, communication, and marketing. This may be explained by Stavanger's engineering culture and the ambitions to develop away from oil dependency (Haarstad, 2016; see also Aunemo et al., chapter 1). In this connection, the establishment of smart city Stavanger may be seen as prompted by external pressures, such as the city's climate and sustainability obligations (Norwegian Ministry of Foreign Affairs, 2016), the EU project Triangulum in 2014, and Nordic Edge Expo in 2015 (Christensen et al., 2015; Evans et al., 2021). Stavanger chose the SC concept as a strategy to tackle its complex issues and foster innovation (cf. Clement et al., 2023; Haarstad, 2016), and thus the SC office aided Stavanger Municipality's development. This 'myth' finds backing in the natural perspective (Christensen et al., 2015).

However, the SC office had been established with employees who were not familiar with Stavanger and had limited cultural understanding of the organization they were employed in, except for the communication employee and the leader, who had previous experience from (cooperation with) Stavanger Municipality. The findings indicate that the recruitment appeared to be a conscious strategy that would give room for recreation of desired practices translated from outside, to promote radical transformation (Røvik, 2016).

> At the beginning, this was very well thought out, we wanted something different. So I got an order from the director, try to bring some new blood into the municipality. The advantage is that we have got employees who naturally think differently. … But at the same time, it is a disadvantage because we have to relate to other municipal employees.
>
> (Leader at the SC office)

After 6 months of operation, the SC office appeared to function effectively according to its mandate, both regarding competence adaptions to tasks/delivery and strategy/direction and regarding the original simplified *smart city* idea with its focus on new technology, collaboration, and citizen involvement.

> Exciting time at the start. Great zeal and will. The employees were professionally skilled in their field and had quite a lot implementation power, and the work was structured.
>
> (Employee at the SC office)

Despite potential conflicts, like insufficient local and cultural knowledge and values of the people in the SC office, these did not lead to a rejection of the *smart city* ideas, as hypothesized in the natural perspective (Christensen et al., 2015; Scott & Davis, 2000; Selznick, 1996). Also, the bottom-up perspective (Easterly, 2008) partially failed (Table 10.1), as there were signs of insufficient anchoring at the grassroots level. In autumn 2019, some resistance to the *smart city* concept was reflected by a chronicle in the local newspaper *Stavanger Aftenblad* (Broch Moe, 2019) that expressed concern that enthusiasm for digital technology may overshadow challenges related to democracy and autonomy. It was met by a replying chronicle with rational arguments in the same newspaper a month later (Crawford, 2020) that pointed out that the smart city of Stavanger is about collaboration and involvement of citizens and businesses. Findings from the interviews underlined that the smart city management had sufficient authority to harmonize such voices threatening the *smart city* idea and to maintain a steady course through established measures (Røvik, 2016), consistent with the rational perspective.

> Yes, we wouldn't be where we are if it wasn't for a leadership that looks a bit up and forward.
>
> (Leader of the SC office)

The *smart city* ideas were expected to be introduced by municipal leaders and employees who developed and implemented the roadmap (Stavanger Municipality, 2016), even if input from multiple stakeholders was sought. Departments began to cooperate and change working methods because of the influence of the SC office.

An inherent culture of change developed through Stavanger Municipality's history of changes and reorganizations may have contributed to a greater understanding of what works and does not work in the municipality and thus allowed new initiatives to be accepted in the organization, in line with the natural perspective (Scott & Davis, 2000). However, the findings also revealed indications that certain parts of the technical environment in Stavanger Municipality had a culture that supports repulsion, while the health department had experience from previous failed implementations.

> If there was something that was really a big gain, it was that the SC office got people around the table, this helped to break down silos. We got people to sit down and talk together, people who had not talked together before, but who should have talked together.
>
> (Employee at the SC office)

> It's not as easy to achieve technological willingness to change among some of the established technical environments. They can and they are positive to participate in experimentation, but they are not as positive to participate in changing work processes. At least, that is our experience so far.
>
> (Head of the SC office)

In parallel with the hard dimension trying to take hold through many smart city projects, the soft dimension (cf. Wataya & Shaw, 2019) appeared. At the leader level, this happened partly through the interpretation of *smart city* ideas in a Norwegian context through the development of a national roadmap for smart and sustainable cities and communities (DOGA, 2019). Smart city Stavanger was strongly involved in the roadmap's development and thus also interested in its implementation. This roadmap has an even greater focus on the soft dimension than the local roadmap (Stavanger Municipality, 2016), including innovative working methods, collaboration across silos both internally and externally, co-creation, and ecological and social sustainability.

> Yes, but the places where it has worked best are the soft places, for example, all the work around co-creation. I won't say it has been incredibly fast, but in a municipal context, the processes have been quite smooth. Here we have got things started and then we have pulled out afterwards. Perhaps the most successful handover we have had is the one we have had with the co-creation school, where the university takes that part from the fall. Then we have taken one out completely and then we have handed it over to external parties.
>
> (Leader of the SB office)

The findings indicated a strong will to advance the SC office and the projects that were anchored in the roadmap (Stavanger Municipality, 2016). However, only a few of these projects were put into use, while established working methods within the municipality took precedence, a phenomenon called *decoupling* in the new institutional perspective (Powell & DiMaggio, 1991; Røvik, 2016). Even though smart city projects might appear in use, there were indications that the introduction of *smart city* as a method stagnated due to a lack of desire and ability to change established environments in Stavanger.

The semi-annual report from smart city Stavanger from spring 2019 (Stavanger Municipality, 2019b) states that the SC office focuses on co-creation and citizen involvement, which highlights the soft dimension of the smart city idea. The report emphasizes that Stavanger has become even better at listening to the citizens in the projects. It also states that the organization had to change not only the way it works but also the organizational culture, while it has succeeded in getting both internal and external environments to work together and thus discovered other ways of working.

However, no signs were found that indicate lasting changes in the departments the SC office collaborated with. This may have been because managers must live up to contradictory demands, including both openness to new ideas and ways of working, such as co-creation and citizen involvement and the delivery of efficiency and quality. Thus, leaders may accept ideas but may not turn them into practice, reflecting decoupling (Meyer & Rowan, 1977; Røvik, 2016), see Table 10.1, assumption 3.

Two things … explain why the soft part goes better. … In the context that it is new and perceived as useful, everyone understands that it makes sense, it is like the tooth of time, co-creation, increased democratization, so there is high awareness around it. On the other hand, the technical part, I think those organizations are so traditional.

(Leader of the SC office)

This situation may reflect the 'garbage can' theory, where solutions often seek problems (Cohen et al., 1972; Røvik, 2016), instead of a purposeful mapping of needs and targeted solutions.

I am so convinced that it is right, and generally for the whole spirit of the time, then in the smart city world. … It is very difficult to create a smart city if you do not do it based on the needs. It is very dumb to create good solutions that there is no need for.

(Employee at the SC office)

Viewed through a new institutional lens (Powell & DiMaggio, 1991), the *smart city* idea appeared as an international trend introduced by city politicians to satisfy the surroundings rather than to put into use and control internal activities.

After two years of operation, the SC office seemed to show transformation, having increased cultural and internal knowledge and more cooperation with other departments in the municipality. The findings revealed that the SC office's work on communication and marketing, both internally and externally, seemed to have developed the SC office employees' cultural understanding related to decoupling (Powell & DiMaggio, 1991) and translation (Czarniawska & Joerges, 1996); see Table 10.1.

Some of the findings underscored tendencies of repulsion as there were further challenges related to the implementation of the bottom-up perspective in Stavanger Municipality. Several statements from the interviews claimed that the SC office had problems with the translation work, due to lack of structure and willingness to follow the smart city method. Also, it was emphasized that several employees at the SC office lacked experience from the public sector and Stavanger Municipality. According to Røvik (2016, p. 299), such a situation can arise when the 'translator' comes from outside and does not have sufficient insider knowledge (Czarniawska & Joerges, 1996).

Most employees lack knowledge of public administration, perhaps most of all, experience.

(Employee at the SC office)

We created an action plan in the first half of the year. This was not a particularly good process, so we worked on it, but then it was not properly revised, nor did it have any consequence. It became more of a wish list rather than a tool to help you prioritize. So there was no formal institutional management of the office. The office lacked structure and formality, and the management did not perceive that perhaps some structure is needed to unleash creativity.

(Employee at the SC office)

On the other hand, the interviews revealed a steady belief in Stavanger Municipality's inherent ability to change and that the SC office represented something different that initially changed the organization. To try to prevent repulsion, the SC office used communication, internal and external marketing, and active network to strengthen the SC office employees' cultural understanding and knowledge of the organization's experience-based knowledge (Christensen et al., 2015). However, no evidence could be found that these measures were sufficient, although it could have been possible for the employees in a newly established office to build up enough insider knowledge to avoid repulsion.

Instead of ending projects, new projects were started. So they were automatically downgraded, but they were not finished. And not evaluated, that's how it was and projects that should have been stopped much earlier.

(Employee at the SB office)

We should have understood why these stranded, and why didn't we succeed? We don't talk about this, at least not towards me.

(Head of a technical department)

The overall findings provided explanatory power to claim that, in fall 2019, the SC office had been through experiences of both quick connection, repulsion, and decoupling (cf. Røvik, 2016; Scott & Davis, 2000; Table 10.1). The semi-annual report of fall 2019 (Stavanger Municipality, 2019a) confirmed that the SC office's work methods and the development of projects had met challenges and announced improvements in the implementation of the smart city method, including a clearer project selection and more straightforward processes for projects to be handed over to operations.

Presumably, after two years of operation, professionally skilled and engaged leaders and employees have developed what is referred to in the translation theoretical perspective as *translator competence* (Czarniawska & Joerges, 1996; Røvik, 2016; cf. Asker Guenduez & Mergel, 2022; Mendoza Moheno et al., 2017). When such competence is developed, the work with both methods and projects will be adjusted to local contexts, including the adaption, adding, emitting, and mixing of elements. To succeed with translation work and contextualization, development arenas are often established (Røvik, 2016). This study's findings confirm that this was done by the smart city Stavanger through the establishment of various networks in the region.

In smart city Stavanger, the implemented ideas, presumably originating both from within and outside the organization, were adjusted and reshaped by local knowledge environments with insights from the EU project Triangulum and local change history (Evans et al., 2021; Haarstad, 2016; Lebrument & La Robertie,

2020). The *smart city* concept, comprising new technology, citizen involvement, and collaborative networks, was obviously influenced by local, regional, national, and international knowledge networks and organizations. Furthermore, private and public companies and academia in the Stavanger area incorporated SC into their portfolios. With initiatives like the Nordic Edge Expo and the Nordic Edge cluster, an emerging smart city cluster has formed in the region, encompassing regional associations, businesses, and academia.

Conclusion

The study indicates that *smart city* ideas as innovative ways of working have not really been established in the organization of Stavanger Municipality. In its first phase, the SC office managed to attract a lot of attention; however, it was mainly disconnected from the culture in Stavanger Municipality. The implementation of *smart city* ideas, described in the local Stavanger roadmap from 2016 (Stavanger Municipality, 2016) and in the national roadmap from 2019 (DOGA, 2019), proved to be challenging due to limited mapping and lack of insight work at the start of the related projects, as well as lack of evaluation of those projects which the SC office had not succeed with.

Stavanger Municipality has been through experiences of quick connection, repulsion, decoupling, and adjusting translation, in view of organizational change theory, dependent greatly on the period under evaluation. However, neither the *smart city* ideas nor the way the SC office worked have taken hold in Stavanger Municipality, and the establishment of the SC office did not lead to profound changes within the organization. Nevertheless, the *smart city* ideas' soft dimensions gained increasing focus, and new working methods have been initiated, reflecting the beginning of a changed way of working.

After two years of operation, leaders and employees obviously have developed translation competences and thus adjusted the *smart city* ideas, work methods, and projects to local contexts; for example, by changes in methods and process and an increased awareness around project selection and their handover to operations. This may help Stavanger Municipality in the future to put out *smart city* ideas as a naturally integrated part of the municipality's work processes.

This study reveals challenges related to the implementation of *smart city* ideas in a medium-sized Norwegian municipality organization in practice. Stavanger Municipality's experiences may be useful for other cities and communities.

Afterwords

In autumn 2023, the SC office in Stavanger lost the status of an independent unit in the administration and is now organized under the leader of Stavanger's innovation and digitalization department (Undheim, 2023). Thus, the SC office's free role – including direct reporting to politicians – is a thing of the past. This means decoupling and a kind of restoring of the past order. The question is whether and how the inspiration and learning from the implementation can be maintained.

One might ponder about the further development of Stavanger Municipality after the merger of the smart city department. Will it be a future-oriented and 'smart' organization considering Stavanger Municipality as a regional engine that needs to reinforce the *smart city* ideas and innovative tools for change work, not least within sustainability?

Acknowledgement

The authors thank Professor Kjell Arne Røvik and Gunnar Edvin Crawford, the head of the smart city office in Stavanger during the time of fieldwork, and his team. Thanks also to the Stavanger Municipality, to the informants, and to the anonymous reviewers.

References

Aase, T.H., & Fossåskaret, E. (2014). *Created realities: On the production and interpretation of qualitative data.* Universitetsforlaget.

Agenda Kaupang. (2018). *Smarte byer og kommuner i Norge – En kartlegging* [Smart cities and municipalities in Norway – A mapping]. Ministry of Local Government and Modernisation. https://www.agendakaupang.no/publication/smarte-byer-og-kommuner-i-norge-en-kartlegging/

Albino, V., Berardi, U., & Dangelico, R.M. (2015). Smart cities: Definitions, dimensions, performance and initiatives. *Journal of Urban Technology*, 22(1), 3–21. doi:10.1080/10630732.2014.942092

Asker Guenduez, A., & Mergel, I. (2022). The role of dynamic managerial capabilities and organizational readiness in smart city transformation. *Cities*, 129, 103791. doi:10.1016/j.cities.2022.103791

Blaikie, N. (2010). *Designing social research: The logic of anticipation.* Polity Press.

Broch Moe, T. (2019, December 15). Smartby Stavanger uten retning [Smart city Stavanger without a direction]. *Stavanger Aftenblad.* https://www.aftenbladet.no/meninger/debatt/i/dO0db1/smartby-stavanger-uten-retning

Chase, S.E. (2013). Narrative inquiry: Still a field in the making. In N.K. Denzin & Y.S. Lincoln (Eds.), *Collecting and interpreting qualitative materials* (4th ed.). SAGE.

Christensen, T., Egeberg, M., Lægreid, P., Roness. P.G., & Røvik, K.A. (2015). *Organizational theory for the public sector* (3rd ed.). Universitetsforlaget.

Clement, J., Ruysschaert, B., & Crutzen, N. (2023). Smart city strategies – A driver for the localization of the sustainable development goals? *Ecological Economics*, 213, 107941. doi:10.1016/j.ecolecon.2023.107941

Cohen, M.D., March, J.G., & Olsen, J.P. (1972). A garbage can model of organization choice. *Administrative Science Quarterly*, 17(1), 1–25. doi:10.2307/2392088

Crawford, G. (2020, January 16). Smartbyen handler om mennesker. *Stavanger Aftenblad.* https://www.aftenbladet.no/meninger/debatt/i/awB1ra/Smartbyen-handler-om-menneske

Czarniawska, B., & Joerges, B. (1996). Travels of ideas. In B. Czarniawska & G. Sevón (Eds.), *Translating organizational change* (pp. 13–48). Walter de Gruyter.

DOGA. (2019). *Roadmap for smart and sustainable cities and communities in Norway. A guide for local and regional authorities developed by Design and Architecture Norway*

(DOGA), the Norwegian Smart City Network and Nordic Edge. https://nordicedge.org/projects/national-smart-city-roadmap/

Easterly, W. (2008). Institutions: Top down or bottom up? *American Economic Review*, 98 (2), 95–99. doi:10.1257/aer.98.2.95

Evans, J., Vacha, T., Kok, H., & Watson, K. (2021). How cities learn: From experimentation to transformation. *Urban Planning*, 6(1), 171–182. doi:10.17645/up.v6i1.3545

Giffinger, R., Fertner, C., Kramar, H., Kakasek, R., Pichler-Milanovic. N., & Meijers, E.J. (2007). *Smart cities. Ranking of European medium-sized cities. Final report.* Center of Regional Science, Vienna University of Technology. doi:10.34726/3565

Gouldner, A.W. (1959). Organizational analysis. In R.K. Merton, L. Broom, & L.S. Cottrell (Eds.), *Sociology today: Problems and prospects* (Vol. 2, pp. 400–428). Harper Torchbook.

Haarstad, H. (2016). Who is driving the 'smart city' agenda? Assessing smartness as a governance strategy for cities in Europe. In A. Jones, P. Ström, B. Hermelin, & G. Rusten (Eds.), *Services and the green economy* (pp. 199–218). Palgrave Macmillan.

Kangas, O., & Kvist, J. (2019). Nordic welfare states. In B. Greve (Ed.), *Routledge handbook of the welfare state* (2nd ed.). Routledge.

Kvale, S. (1997). *Det kvalitative forskningsintervju*. Gyldendal.

Latour, B. (1987). *Science in action: How to follow scientists and engineers through society.* Harvard University Press.

Lebrument, N., & de La Robertie, C. (2020). Unplugged – Thinking the organizational and managerial challenges of intelligent towns and cities: A critical approach to the smart city phenomenon. *M@n@gement*, 22(2), 357–372.

Meijer, A., & Bolivar, M.P.R. (2016). Governing the smart city: A review of the literature in smart urban governance. *Internation Review of Administrative Sciences*, 82(2), 392–408. doi:10.1177/0020852314564308

Mendoza Moheno, J., Hernandez Calzada, M.A., & Salazar Hernandez, B.C. (2017). Organizational challenges for building smart cities. In M. Peris-Ortiz, D. Bennett, & D. Pérez-Bustamante Yabar (Eds.), *Sustainable smart cities. Innovation, technology, and knowledge management* (pp. 89–99). Springer. doi:10.1007/978-3-319-40895-8_7

Meyer, J., & Rowan, B. (1977). Institutional organizations: Formal structure as myth and ceremony. *American Journal of Sociology*, 83(2), 340–363.

Norwegian Ministry of Foreign Affairs. (2016). *Norway's follow-up of agenda 2030 and the Sustainable Development Goals.* www.regjeringen.no/en/dokumenter/follow-up-sdg2/id2507259

Powell, W.W., & DiMaggio, P.J. (Eds.). (1991). *The new institutionalism in organizational analysis.* University of Chicago Press.

Røvik, K.A. (2016). Knowledge transfer as translation: Review and elements of an instrumental theory. *International Journal of Management Reviews*, 18, 290–310. doi:10.1111/ijmr.12097

Scott, R.W., & Davis, G.F. (2000). *Organizations and organizing: Rational, natural, and open systems perspectives* (1st ed.). Routledge. doi:104324/9781315663371

Selznick, P. (1996). Institutionalism 'old' and 'new'. *Administrative Science Quarterly*, 41 (2), 270.

Stavanger Municipality. (2016). *Roadmap for the smart city Stavanger – Vision, goals and priority areas.* https://www.stavanger.kommune.no/siteassets/samfunnsutvikling/planer/engelske-planer/roadmap-smart-city-stavanger-2016.pdf

Stavanger Municipality. (2019a). *Smartbyen Stavanger – høst 2019. Halvårsrapport (half-year report autumn)*. https://www.stavanger.kommune.no/samfunnsutvikling/smartbyen-stavanger/halvarsrapport/host19

Stavanger Municipality. (2019b). *Smartbyen Stavanger – våren 2019. Halvårsrapport (half-year report spring)*. https://www.stavanger.kommune.no/samfunnsutvikling/smartbyen-stavanger/halvarsrapport/Vaar19/

Thomas, G. (2011). A typology for the case study in social science following a review of definition, discourse, and structure. *Qualitative Inquiry*, 17(6), 511–521. doi:10.1177/1077800411409884

Undheim, S. (2020). *Implementation of the smart city concept in Stavanger Municipality: A case study of how Smart City Stavanger and selected measures meet the field of practice* [Master's thesis, University of Stavanger]. https://uis.brage.unit.no/uis-xmlui/handle/11250/2682071

Undheim, S. (2023, December 6). Hva nå, smart city? *Stavanger Aftenblad*. https://www.aftenbladet.no/meninger/debatt/i/EQ1vqA/hva-naa-smart-city

Vedeld, T. (2022). The co-creation paradox: Small towns and the promise and limits of collaborative governance for low-carbon, sustainable futures. *Scandinavian Journal of Public Administration*, 26(3), 45–70. https://publicera.kb.se/sjpa/issue/view/781

Wataya, E., & Shaw, R. (2019). Measuring the value and the role of soft assets in smart city development. *Cities*, 94, 106–115. doi:10.1016/j.cities.2019.04.019

Weber, M. (1968). *Economy and society* (G. Roth & C. Wittich, Eds. & Trans.), Bedminster Press. (Original work published 1922)

Zey, M.A. (2015). Rational choice and organization theory. In N.J. Smelser & P.B. Baltees (Eds.), *International encyclopedia of the social and behavioral science* (2nd ed., Vol. 19, pp. 892–895). Elsevier. doi:10.1016/B978-0-08-097086-8.73109-6

11 Enabling the future smart cities

AI-based orchestration of 5G and beyond

Gianfranco Nencioni, Annisa Sarah and Anders Riel Müller

Introduction

Make a Google image search on smart cities. Most of the images you will see show a city with ubiquitous wireless connections tying together buildings, cars, people, and so forth. It is perhaps coincidence that the popularity of the smart city as a concept has grown in parallel with the spread of wireless communication networks. The smart city and wireless communications are intimately tied to each other. Yet the techno-optimistic predictions of the connected city where human life, energy, transport, and energy flow seamlessly throughout the city, managed and enabled by powerful computers, algorithms, and wireless communication technologies, have yet to materialize. While some solutions have been implemented, the predicted revolution of urban life is far from the futuristic images of smart cities that Google images suggest (Clark, 2021; Yigitcanlar et al., 2019).

Wireless communications technologies play a pivotal role in delivering smart city services (Yaqoob et al., 2017). Such technology is expected to allow faster and more efficient data exchange and management, resulting in faster operation. Hundreds to thousands of sensors send a vast amount of data; a high-computing server then processes the data and sends back the response the users need. With a huge amount of data, artificial intelligence (AI) can process the data, and this allows for the automation of city operations, ensuring connected processes that are cost-effective and contribute to energy conservation and overall sustainability (Yigitcanlar et al., 2021). So far these promises of new and optimized cities have run into several stumbling blocks including lack of investments in infrastructure, technical challenges, and societal opposition.

Technical challenges undermining current smart cities

Most smart city services now use LoRaWAN or 4G (LTE-M), which are long-range technologies for Internet of Things (IoT; Adelantado et al., 2017). *Long-range* means that an access point or base station can cover a long (kilometres) distance. *IoT* refers to a collection of many interconnected sensors and processing devices. Due to the limited capacity of these technologies, there has been a struggle to handle the sheer volume of devices and data required for widespread smart city

DOI: 10.4324/9781003498650-15

applications (Adelantado et al., 2017). Furthermore, such technologies cannot provide the low-latency connections that are required for more demanding real-time applications such as autonomous vehicles or emergency response systems (Gohar & Nencioni, 2021b). These technologies are also struggling to provide the data rate for high-definition video streaming, remote healthcare services, and other data-intensive and latency-sensitive applications (Astound, 2023). These performance limitations can be overcome by the fifth generation (5G) of mobile networks, which enhances the capacity, lowers the latency, and improves the reliability. These 5G characteristics may also enable new innovative smart city services and applications.

If 5G can address these technological challenges, other techno-economic and socio-economic challenges still need to be addressed (see, for example, Huang et al., chapter 3, Lindland, chapter 5, and Fisker et al., chapter 6). The first step to providing smart city services is building the communication infrastructure and customizing it as needed (Sarah et al., 2023). The costs for infrastructure and hiring relevant engineers with new technical skills and knowledge are high and entail further challenges. The 5G network is expected to utilize infrastructure softwarization and virtualization to enable high flexibility and customization to fulfil different service requirements (Nencioni et al., 2018). Virtualization may suppress capital expenditure and allow business actors to enter the market more efficiently.

Softwarization and virtualization bring sustainability because, instead of having a dedicated device that needs to be substituted when a new generation comes, the functionality in a virtual environment is just software running software on top of general-purpose hardware, which just needs to be upgraded to move to a new mobile generation. However, how these technologies will be able to provide secure and dependable services is still an open question. Moreover, the energy consumption of such technologies may have a huge impact. For example, in Italy, Telecom Italia has been the second largest consumer of electricity after the National Railway (Pileri, 2007).

Challenges related to wireless communications for smart city services

Our point of departure is our experience from Stavanger, where a test arena for autonomous mobility and transport was sought and established. In 2017–2018, the local company Mobility Forus, the regional transit authority Kolumbus, and Forus Business Park began trials for self-driving buses in Forus, Norway's largest industrial park. The aim of the trials was to examine the use of these buses to better connect parts of the business park that were the farthest away from major bus lines. This was particularly relevant for the parent company of Mobility Forus, Seabrokers, who owns several office complexes in that part of Forus. The vision was that the last mile from the closest bus to the office complexes could be serviced by these self-driving buses.

After the initial tests in 2017–2018, Forus Mobility together with Forus Business Park proposed a plan to use an entire district of the business park as national and international test arena for autonomous mobility and transport and the establishment of the YAGO business cluster that included the University of Stavanger (UiS) among others as partners (Yago, n.d.). The test arena faced numerous challenges from the beginning. A lack of funding to invest in the necessary technical infrastructure for next-generation mobility solutions was the biggest hurdle. The first self-driving bus was a GPS-enabled unit that was pre-programmed to run on a specific route (Kolumbus, 2018). The next generation of autonomous solutions requires an entirely new wireless communication and computing infrastructure including 5G and multi-access edge computing (MEC; Nencioni et al., 2023).

Challenges on retrieving funding

There were attempts to raise funding to invest in a limited private 5G network, sensors, and MEC hardware on a stretch of road in Forus, but funding was never granted. Such infrastructure would have fitted well with an ongoing project at UiS, 5G-MODaNeI, that sought to provide a secure and dependable resource orchestration of 5G and MEC infrastructure. The project focuses on developing effective and intelligent algorithms for the resource orchestration and identifying the main security and dependability challenges for 5G-MEC systems. The security and dependability of network infrastructure and services are necessary, primarily to deliver mission-critical services such as public safety scenarios. Although a few attempts were made to raise funds for a limited 5G-MEC test infrastructure in Forus, there seemed to be a lack of willingness to invest from private funders. It appears that the vision to establish a test arena for autonomous mobility and transport in Forus has been put on hold for now.

The case of the Forus Test arena and the Yago autonomous technology cluster provides an example of a vision that lacked both technical and financial capacity when it comes to meeting the requirements of a 5G test arena. Even with those investments realized, many technical challenges for smart city services that rely on 5G and MEC remain. Unlike the optimistic scenarios of served by ubiquitous wireless networks and powerful computers, the reality is still far away from realizing the full potential of 5G. This chapter draws on the findings and insights of the 5G-MODaNeI project at UiS.

Chapter contribution and outline

The contribution of this chapter is to identify the technical background of the unmet promise of a smart city based on wireless communication, what is currently lacking in terms of technology, and what issues need to be addressed to enable the future smart city and how they can be overcome. The technical background of 5G, MEC, and relevant technologies to enable smart city solutions is provided. Then, a description of the 5G-MODaNeI project is presented. Challenges and opportunities for future smart cities are discussed, followed by recommendations.

5G mobile communications

A mobile network is composed of the following architectural elements: the air interface, which is the wireless connection between the user equipment and the base station; the core network, which manages the mobile network and provides networking services to the access network; and the radio access network, which includes the base station and the connectivity to the core network. Most of the technical innovations that enable improved performance are at the air interface (Stallings, 2021).

Mobile networks have developed quickly in the past decades (Stallings, 2021). The first generation (1G) was an analogical technology and consisted only of voice calls. The second generation (2G) was digital and included Short Messaging Service (SMS). The third generation (3G) was the first generation to natively include Internet connectivity. The fourth generation (4G) introduced actual broadband connectivity. Two versions of 4G, called NarrowBand-IoT (NB-IoT) and Long-Term Evolution for Machines (LTE-M), have been developed to support long-range IoT communications. 5G is enhancing the 4G services by supporting enhanced Mobile Broadband (eMBB) and massive Machine-Type Communication (mMTC) and enabling new kinds of services, like Ultra Reliable Low-Latency Communication (URLLC; Stallings, 2021).

The key innovation of the 5G radio access network and core network is the softwarization and virtualization, which include networking paradigms such as software-defined networking (SDN) and network function virtualization (NFV; Nencioni et al., 2018). SDN and NFV make it possible to use commercial off-the-shelf hardware instead of dedicated hardware. We do not need a specific device to implement network functionality – just software that can be installed anywhere. Moreover, there is not even a dedicated physical device for a network, but the hardware resources are virtualized. This is like bringing the concept of cloud computing to the communication networks (Nencioni et al., 2023).

SDN and NFV are also key enablers of network slicing, which consists of realizing virtual networks with heterogenous requirements on top of a shared infrastructure (Nencioni et al., 2018). This makes it possible to provide all 5G services (eMBB, mMTC, and URLLC). MEC, which consists of providing computing capability close to the users (different from data centres, which are somewhere most likely far away), is one of the enablers of the URLLC. MEC not only reduces the latency but also enables context real-time awareness, cloud offloading, the reduction of traffic congestion, and the improvement of data privacy (Nencioni et al., 2023).

The 5G that is currently deployed by network operators across the world is, however, technically not full 5G (Ericsson, 2022). To realize a smooth transition from 4G to 5G, migration strategies have been implemented where a mobile network, for example, has a 5G radio access network and a 4G core network (or vice versa). This version of 5G is called Not Stand Alone (NSA). The Stand-Alone version is currently under deployment (GSMA, 2019a). Moreover, 5G standardization is not something static. The first release was in 2019, two specification releases followed, and in 2024 there was another major release defining 5G Advanced. The development of the 'true 5G' is thus an incremental affair.

One of the possible deployments of true 5G is private 5G, which is a non-public network. This deployment option, which is also possible in other generations, has not been extensively used so far but is expected to gain success, especially in what is called the fourth industrial revolution or Industry 4.0 (Flō Networks; 2023). Private 5G consists of small-size networks, located at the use case premises and locally administrated. Private 5G connects only a reduced number of devices and provides a customized service for specific use case needs. Private 5G helps to enable low latency, a high data rate, and seamless secure wireless connectivity.

6G and AI

Even if the real potential of 5G is far from being realized, the standardization of 6G has started. The general definition of the 6G framework has been recently released by the International Telecommunication Union (2023). The current 6G vision includes an extension of 5G usage scenarios: from eMBB to immersive communication, from mMTC to massive communication, from URLLC to hyper-reliable low-latency communication. New usage scenarios, such as ubiquitous connectivity, AI and communication, integrated sensing and communication, are also envisioned. The general 6G principles are sustainability, connecting the unconnected, ubiquitous intelligence, and security and resilience. One of the key aspects of 6G will be the native use of AI by any architectural element. Even if it was not natively envisaged, AI is already extensively used by 5G (Sarah et al., 2023). One context in which the use of AI is essential is in orchestration and management.

The 5G-MODaNeI project

5G-MODaNeI is a project funded by the Norwegian Research Council, led by the Department of Electrical Engineering and Computer Science at UiS and in collaboration with the University of Pisa (Italy) and the University of Bucharest (Romania). The focus is on security and dependability in the orchestration of data and network resources in 5G-MEC systems. 5G-MODaNeI consists of three main parts. The first part focuses on identifying the risks and vulnerabilities due to attacks and failures in 5G-MEC systems. Initially, the 5G-MEC architecture was studied and the state of the art regarding security, dependability, and performance was surveyed. The challenges of jointly addressing the three aspects have been discussed (Nencioni et al., 2023). A first evaluation of the dependability of a 5G-MEC system was performed by proposing a model of the system availability; that is, the readiness to provide a service with the desired requirements (Pathirana & Nencioni, 2023).

The second part focuses on the development of innovative and intelligent solutions to allocate the network and data resources for maximizing security and dependability in 5G-MEC systems. Firstly, the state of the art on resource allocation in 5G-MEC systems was surveyed (Sarah et al., 2023). Then, resource allocation from security and dependability perspectives was investigated (Khan &

Nencioni, 2023a). Novel solutions to allocate data and network resources in a 5G-MEC scenario have been proposed. These solutions are based on various techniques: heuristic (Gohar & Nencioni, 2021a), mathematical optimization (Sarah & Nencioni, 2024), and machine learning (ML; Mason et al., 2023; Sarah et al., 2024). Furthermore, some works addressed the security and dependability of ML-based solutions, which may be attacked by adversaries (Khan & Nencioni, 2022, 2024) or subjected to failures (Khan & Nencioni, 2023b). Another work proposed an ML-based solution for the MEC application migration; that is, moving the MEC application from one MEC server to another (Sarah et al., 2024).

The third part focuses on the realization of a 5G-MEC experiment in a vehicular scenario to test the proposed solutions. Firstly, the state of the art on 5G testbeds in vehicular scenarios was surveyed (Wadatkar et al., 2023a). The MEC application migration is due to the mobility of the user or the failure of the MEC server where the MEC application is running. A hybrid 5G-MEC testbed has been developed (Wadatkar et al., 2022). The testbed integrates various 5G and MEC simulators and emulators, includes actual MEC servers, and proposes a controller. The controller allows the experimentation on MEC application migration. The controller interacts with the MEC framework by using standard MEC interfaces and allows the use of complex migration solutions (Wadatkar et al., 2023b, 2024).

Moreover, different migration strategies have been compared (Hathibelagal et al., 2023) and a migration solution that allows aiming at multiple targets has been evaluated (Garroppo et al., 2022).

Moving towards the future smart cities

5G – what do we have now and what is lacking?

Current pre-5G technologies, such as LoRaWAN and LTE-M, have been struggling to deliver smart city solutions due to their limited bandwidth and capacity. This limitation can hinder real-time data transmission and processing for large-scale deployments. Future network technologies must be customizable so that we can have high-resolution video streaming with a very low latency. Current smart city services, such as sewage monitoring and waste management, usually have rigid and fragmented infrastructure and organizational boundaries, meaning there is limited collaboration or resource sharing that may benefit users or service providers (Montori et al., 2017). A lack of collaboration between parties and providers is usually rooted in privacy and security concerns (see Skotnes and Gould, chapter 8). Municipalities must navigate complex privacy and security regulations (for example, General Data Protection Regulation), while the infrastructure provider and other active business actors must ensure the security requirements will be fulfilled from end-to-end services (Braun at al., 2018). Moreover, a comprehensive integration solution requires significant investment in both technology infrastructure and human resources, which may not be feasible as long as the business models for 5G remain uncertain (Banda et al., 2022). Who will invest, who will manage, who is willing to pay, and what are the regulatory frameworks that allocate responsibilities to different parties? These questions remain unsolved. There is a need for communication technologies with better performance that allow collaboration and data/resource sharing amongst smart city actors, but we also need the organizational setup that allows collaboration and that is economically sustainable for the actors involved.

Discussion: why are 5G and MEC central to future smart city solutions?

As true 5G enables services with very different requirements in the same physical infrastructures, the infrastructure will be highly customizable due to softwarization and virtualization. MEC provides resources in the edge, which may be used as a server to process computation-hungry applications such as AI video analytics. The proximity of the resources to the users reduces the latency, leading to faster response times between applications and users and enabling URLLC services. The proximity of the resources to the users also reduces the exposition of the data traffic, leading to an enhancement of the privacy. 5G-MEC systems can therefore provide the performance required by smart city services that current technology cannot.

When true 5G is rolled out, the softwarized and virtualized resources are expected to be agile in size and location and have a standard protocol to communicate between different operators or parties (ETSI, 2022). The open and standardized communication protocols may allow collaboration between parties and the sharing of relevant data to achieve efficient and optimized services to the end users. Consequently, 5G-MEC systems are expected to be able to solve the fragmented infrastructure issue by having a highly configurable, software-based, virtualized network infrastructure. The introduction of true 5G will also happen depending on the economic interest of network operators and the demand by other smart city actors. For example, the commercialization of mMTC most probably happens when network operators have obtained a profitable return on investment on 4G solutions; that is, NB-IoT and LTE-M.

Currently, there are first efforts to introduce private 5G (Nag, 2022), which are overcoming some of the current limitations by providing a secure (on-premises and locally administrated), fully customized communication infrastructure. In Stavanger, for example, a private 5G network is currently being installed in the new hub for urban innovation Innoasis. Innoasis is a co-working space for small businesses and start-ups working on smart city solutions. Together with the local energy and telecommunications company Lyse, a private 5G network is planned to allow companies to test and demonstrate solutions (Nag, 2022).

Challenges and potential new issues: what needs to be done to complete the puzzle?

In addition to the technical challenges to be solved to unlock the full potential of 5G, social, economic, and environmental issues need to be addressed (Muench et al., 2022). Some of the key challenges identified in the 5G-MODaNeI project are presented here.

Technical, sustainability, techno-economic, and techno-social challenges

The softwarization and virtualization of infrastructure may imply new challenges, like managing unexpected *security* threats and bugs that lead to failures and undermine the *dependability* in a commercial deployment. There are three technical phases in providing 5G-MEC services: infrastructure deployment, network slice management, and end-user service management (Sarah et al., 2023). The first phase includes deploying both physical and virtualization infrastructure. The second phase is creating and managing a network slice. The last phase is allocating and managing a service for the end users within the network slice.

These phases are concerned with not only *network* resources but also *data* resources. Both kinds of resources should be considered in an integrated way to obtain the full potential of 5G-MEC systems.

These considerations highlight how the management and orchestration of 5G-MEC systems are much more *complex* than in previous technologies.

5G-MODaNeI focuses on the last two phases analyzing the behaviours of 5G-MEC systems and proposing effective solutions for management and orchestration. During the deployment phase, the main challenge is to decide the scale, capacity, and coverage with the constraint of cost. Infrastructure providers have a limited amount of money to invest; however, they need to provide satisfactory services to users. This balance between investments and usability is difficult to assess in advance. In addition, the infrastructure provider needs to consider the opportunity of having a backward compatibility solution and having a seamless integration between 5G and other technologies. This phase is mostly dependent on the business strategy of infrastructure providers, which is connected to local business activities.

After the infrastructure is ready, the network slices, created on top of the physical infrastructure, need to be managed. New actors such as slice brokers can enter the market and take the role of network slice configurator and bridge the slice tenants to infrastructure providers. Challenges arise when a slice broker needs to buy resources from several infrastructure providers, serving slice tenants and making a profit (Gohar & Nencioni, 2021b; Sarah & Nencioni, 2024). There are also interoperability issues when we integrate different 5G providers, cloud providers, and MEC providers. Moreover, the isolation of the network slices, in terms of performance, security, and dependability, is a critical aspect that should be considered (Gonzalez et al., 2020).

The network and data resource allocation process can be done by using various techniques. Given the complexity and dynamicity of the environment, the use of AI has obvious benefits. AI algorithms will aim to have performing, secure, and dependable 5G-MEC services, while themselves being targets of security attacks and having dependability issues. These aspects should be better investigated and considered (Khan & Nencioni, 2022, 2023a, 2024). This is also important for future 6G networks, where AI is natively embedded at all levels of the network.

When the 5G infrastructure is well deployed and network slices are available, we must take care of user service performance. When the user is mobile (e.g., a moving car), the service latency may be decreased, and we need to maintain performance and provide high reliability of services as the user moves. From the perspective of infrastructure, service management, and orchestration, including maintaining service performances, several aspects must be addressed such as security and dependability of both system and services. There are threats and failures that may happen unexpectedly. These challenges need to be addressed to maintain safe and satisfactory services, especially for life-threatening use cases such as autonomous driving services.

Sustainability

From a sustainability perspective, the 5G-MEC implementation supports at least two of the Sustainable Development Goals (SDGs; United Nations, 2015). First, the combination of AI and 5G-MEC can build resilient infrastructure and sustainable industrialization, which is referred to in Goal 9 of the SDGs (World Economic Forum, 2020). 5G services have an impact on the manufacturing industry with automation and predictive intelligence of factory processes; for instance, improving more than 40% of the operational efficiency of the Nokia factory in Oulu (Nokia, 2019). Another case uses 5G services to recognize defects in assembly line patterns, which contribute to increased production volume and achieve a return on investment of 29 times over 2 years (DataProphet, n.d.). Second, with AI and 5G-MEC, buildings and cities can be safer and more efficiently managed by implementing a network of sensors and actuators, referred to in Goal 11 of the SDGs (World Economic Forum, 2020). For example, real-time analytics help Turkey with its emergency disaster response (GSMA, 2019b), and IoT services help Croatia and Brazil monitor air quality parameters and pollution levels (Telefonica, 2019).

Technically, softwarization and virtualization technologies have an impact on the lifetime of hardware devices, since new technologies will need only software updates rather than replacing the hardware infrastructure. This will potentially lead to a reduction of costs, waste, and associated energy consumption and pollution. Network slicing by sharing the same infrastructure for multiple use cases is also leading to a reduction of deployed devices and enables a more efficient use of resources. Yet, the impact on direct energy consumption during the operations of 5G-MEC system is still unknown (Ericsson, 2023; STL Partners, n.d.).

Techno-economic and socio-technological challenges

Other societal and economic challenges will require attention even after 5G networks are fully available; for example, techno-economic challenges from an industry, sustainability, and energy perspective, as well as societal aspects of end users. Moreover, since 5G enables new use cases, 6G will enable even more. Because classical broadband user connectivity is not profitable anymore, new applications and services should be created to exploit the new capability and provide new advanced services. This requires infrastructure providers to deploy and commercialize an infrastructure with advanced characteristics. Service providers need to be aware of such technical opportunities and be able to design applications and services. Unfortunately, this is a chicken/egg problem. Network operators are eager to show the potential of 5G technology in relevant use cases, and service providers and other actors are not aware of what 5G is able to realize (Brozynski & Leibowicz, 2022). Network operators are therefore reluctant to invest in true 5G infrastructure until they have a business case, and service providers are not able to understand the business potential of 5G networks until they can test it.

A solution is to create open 5G testbed arenas (Lyse 5G lab at Innoasis, for example), where 5G applications and services can be developed and tested (Nag, 2022; Orange, n.d.). This solution needs a multidisciplinary approach because creating 5G-based smart city applications that revolutionize future smart city services requires both knowledge of 5G-MEC technology and user and citizen needs. These were also the challenges in the plans to develop a 5G test arena for autonomous solutions in the Forus industrial park in Stavanger. Funders were reluctant to invest in 5G infrastructure because it was unclear among service providers how to make a business out of this 5G infrastructure. It may be useful here to provide an example of how this works. The 3G network was introduced in the early 2000s, but it was the introduction of the 3G-enabled iPhone in 2007 that provided a compelling business case for faster and broader 3G network rollout (Akematsu et al., 2012). Thus, as the iPhone became popular, users began to demand faster speeds and better 3G coverage. Such a use case is still missing for 5G.

Conclusions

In this chapter, we have introduced the current situation and limitations of existing smart city solutions, focusing on experiences observed in Stavanger. The chapter identified performance constraints stemming from current communication technologies, the fragmentation of telecommunication infrastructures, security concerns, and substantial investment requirements as key challenges. Novel technologies are currently being developed that are poised to address some of these limitations. For instance, 5G-MEC systems promise enhanced performance by leveraging softwarization and virtualization. These paradigms aim to reduce deployment and operational costs while enabling network slicing, facilitating isolated virtual networks, and accommodating heterogeneous performance demands on top of shared infrastructures.

However, the advent of 5G-MEC systems also poses fresh challenges, spanning technical, societal, and environmental domains. These challenges encompass the complexity inherent in orchestrating a system of systems, ensuring the security and dependability of nascent technological ecosystems, understanding the sustainability implications (such as energy consumption), and navigating the interplay between infrastructure deployment and service development dynamics. Moreover, we explore how ongoing initiatives, such as the 5G-MODaNeI project at UiS and research into 6G, are addressing some of these challenges. Efforts include leveraging AI to manage system complexity, although this introduces additional considerations regarding the security and dependability of AI systems.

True 5G deployments will provide the wireless communications infrastructure that will open possibilities for new smart city solutions to be developed. But use cases and business cases are still too few for network operators to see a return on investment for developing these infrastructures.

In conclusion, we propose recommendations for relevant stakeholders to materialize the vision of future smart cities:

Telecom companies: Invest in and develop 5G infrastructure, fostering innovation and collaboration with diverse industries to tailor 5G applications for specific smart city requirements.

Government entities: Encourage communication infrastructure investment through incentives and partnerships with telecom companies, while providing clear regulations and support to expedite deployment.

Other sectors (industries, academia): Identify mutual benefits of 5G adoption, contribute to standards development, participate in testing 5G smart city applications, and offer feedback on practical implementation experiences.

Private 5G network testbeds may provide a useful entrance into exploring the potential use cases for 5G but require willing investors and cross-sectoral collaboration at a deeper level than what has been observed in Stavanger until now.

Funding

This work was partially supported by the Norwegian Research Council through the 5G-MODaNeI project (No. 308909).

References

Adelantado, F., Vilajosana, X., Tuset-Peiro, P., Martinez, B., Melia-Segui, J., & Watteyne, T. (2017). Understanding the limits of LoRaWAN. *IEEE Communications Magazine*, 55(9), 34–40.

Akematsu, Y., Shinohara, S., & Tsuji, M. (2012). Empirical analysis of factors promoting the Japanese 3G mobile phone. *Services, Regulation and the Changing Structure of Mobile Telecommunication Markets*, 36(3), 175–186.

Astound. (2023). *Data bandwidth: Critical to healthcare providers.* White Paper. https://www.astound.com/business/wp-content/uploads/2023/09/Brainstorm_Astound_Healthcare_White_Paper-2023-08.pdf

Banda, L., Mzyece, M., & Mekuria, F. (2022). 5G business models for mobile network operators – A survey. *IEEE Access*, 10, 94851–94886.

Braun, T., Fung, B.C., Iqbal, F., & Shah, B. (2018). Security and privacy challenges in smart cities. *Sustainable Cities and Society*, 39, 499–507.

Brozynski, M.T., & Leibowicz, B.D. (2022). A multi-level optimization model of infrastructure-dependent technology adoption: Overcoming the chicken-and-egg problem. *European Journal of Operational Research*, 300(2), 755–770.

Clark, J. (2021). What cities need now. *MIT Technology Review*, 124(3), 4–8.

DataProphet. (n.d.). *Robotic stud welding excellence for a premium automotive plant.* https://dataprophet.com/case-study-stud-welding/

Ericsson. (2022). *5G SA deployment: Moving beyond eMBB.* https://www.ericsson.com/49f649/assets/local/reports-papers/mobility-report/documents/2022/063022-emr-june-2022-5g-sa-deployment-article-web.pdf

Ericsson. (2023). *How 5G will drive Jordan's Vision 2025 and sustainable development.* https://www.ericsson.com/en/blog/5/2023/how-5g-will-drive-jordans-vision-2025-and-sustainable-development

ETSI. (2022, June). *ETSI White Paper No. 49: MEC federation: Deployment considerations.*

Flō Networks. (2023). *Revolutionizing Industry 4.0 with private 5G technology and cloud computing.* https://flo.net/revolutionizing-industry-4-0-with-private-5g-technology-and-cloud-computing

Garroppo, R.G., Volpi, M., Nencioni, G., & Wadatkar, P.V. (2022, September 5–8). *Experimental evaluation of handover strategies in 5G-MEC scenario by using AdvantEDGE* [Paper presentation]. IEEE International Mediterranean Conference on Communications and Networking (MeditCom), Athens, Greece.

Gohar, A., & Nencioni, G. (2021a, May 10–13). *Minimizing the cost of 5G network slice broker* [Paper presentation]. IEEE INFOCOM 2021-IEEE Conference on Computer Communications Workshops (INFOCOM WKSHPS), Vancouver, BC, Canada.

Gohar, A., & Nencioni, G. (2021b). The role of 5G technologies in a smart city: The case for intelligent transportation system. *MDPISustainability*, 13(9), 5188.

Gonzalez, A.J., Ordonez-Lucena, J., Helvik, B.E., Nencioni, G., Xie, M., Lopez, D.R., & Grønsund, P. (2020, June 15–18). *The isolation concept in the 5G network slicing* [Paper presentation]. 2020 European Conference on Networks and Communications (EuCNC), Dubrovnik, Croatia.

GSMA. (2019a). *Operator requirements for 5G core connectivity options.* https://www.gsma.com/futurenetworks/wp-content/uploads/2019/05/20190515-GSMA-Operator-Requirements-for-5G-Core-Connectivity-Options.pdf

GSMA. (2019b). *Utilising real-time mobile analytics to inform emergency disaster response in Turkey.* https://www.gsma.com/convening-for-good/resources/utilising-real-time-mobile-analytics-to-inform-emergency-disaster-response-in-turkey

Hathibelagal, M.A., Garroppo, R.G., & Nencioni, G. (2023). Experimental comparison of migration strategies for MEC-assisted 5G-V2X applications. *Computer Communications*, 197, 1–11.

International Telecommunication Union. (2023). *Recommendation ITU-R M.2160–0: Framework and overall objectives of the future development of IMT for 2030 and beyond.*

Khan, M.M.I., & Nencioni, G. (2022, December 18–21). *Revenue-model learning for a slice broker in the presence of adversaries* [Paper presentation]. IEEE International Conference on Advanced Networks and Telecommunications Systems (ANTS), Gandhinagar, Gujarat, India.

Khan, M.M.I., & Nencioni, G. (2023a). Resource allocation in networking and computing systems: A security and dependability perspective. *IEEE Access*, 11, 89433–89454.

Khan, M.M.I., & Nencioni, G. (2023b, June 20–27). *Revenue maximization of a slice broker in the presence of Byzantine faults* [Paper presentation]. 2023 53rd Annual IEEE/IFIP International Conference on Dependable Systems and Networks Workshops (DSN-W), Porto, Portugal.

Khan, M.M.I., & Nencioni, G. (2024). *Adaptive methods for revenue model learning of a slice broker in the presence of adversaries.* Wireless Personal Communications.

Kolumbus. (2018). *Autonomous future.* https://www.kolumbus.no/en/news-archive/autonomous-future-at-forus/

Mason, F., Nencioni, G., & Zanella, A. (2023). Using distributed reinforcement learning for resource orchestration in a network slicing scenario. *IEEE/ACM Transactions on Networking*, 31(1), 88–102.

Montori, F., Bedogni, L., & Bononi, L. (2017). A collaborative Internet of Things architecture for smart cities and environmental monitoring. *IEEE Internet of Things Journal*, 5(2), 592–605.

Muench, S., Stoermer, E., Jensen, K., Asikainen, T., Salvi, M., & Scapolo, F. (2022). *Towards a green and digital future.* Publications Office of the European Union. doi:10.2760/977331

Nag, A.T. (2022). *Er bedriftene klare for 5G?* https://www.naeringsforeningen.no/nyh eter/er-bedriftene-klare-for-5g/

Nencioni, G., Garroppo, R.G., Gonzalez, A.J., Helvik, B.E., & Procissi, G. (2018). Orchestration and control in software-defined 5G networks: Research challenges. *Wireless Communications and Mobile Computing*, 2018.

Nencioni, G., Garroppo, R.G., & Olimid, R.F. (2023). 5G multi-access edge computing: A survey on security, dependability, and performance. *IEEE Access*, 11, 63496–63533.

Nokia. (2019). *Customer case: Nokia Oulu factory.* https://assets.omron.eu/downloads/p ublication/en/v1/nokia-oulu-ld_customer_reference_en.pdf

Orange. (n.d.). *Orange 5G Lab.* https://5glab.orange.com/en/

Pathirana, T., & Nencioni, G. (2023, July 24–27). *Availability model of a 5G-MEC system* [Paper presentation]. 2023 32nd International Conference on Computer Communications and Networks (ICCCN), Honolulu, HI, USA.

Pileri, S. (2007, September 30–October 4). *Energy and communication: Engine of the human progress* [Keynote]. International Communications Energy Conference (INTELEC), Rome, Italy.

Sarah, A., & Nencioni, G. (2024). Resource allocation for cost minimization of a slice broker in a 5G-MEC scenario. *Computer Communications*, 213, 331–344.

Sarah, A., Nencioni, G., & Khan, M.M.I. (2023). Resource allocation in multi-access edge computing for 5G-and-beyond networks. *Computer Networks*, 227, 109720.

Sarah, A., Nencioni, G., & Khan, M.M.I. (2024). DRL-based availability-aware migration of a MEC service. *IEEE Open Journal of the Communications Society*, 5, 5088–5102.

Stallings, W. (2021). *5G wireless: A comprehensive introduction.* Addison Wesley Professional.

STL Partners. (n.d.). *5G and sustainability: The role of green 5G in the energy transition.* https://stlpartners.com/articles/sustainability/5g-and-sustainability/

Telefonica. (2019). *Consolidated management report.* https://www.telefonica.com/en/wp-con tent/uploads/sites/5/2021/08/2019-Telefonica-Consolidated-Management-Report.pdf

United Nations. (2015). *Transforming our world: The 2030 agenda for sustainable development.* https://sdgs.un.org/2030agenda

Wadatkar, P.V., Garroppo, R.G., & Nencioni, G. (2022, November 23–25). *MEC application migration by using AdvantEDGE* [Paper presentation]. EAI International Conference on Tools for Design, Implementation and Verification of Emerging Information Technologies (TRIDENTCOM), Melbourne, Australia.

Wadatkar, P.V., Garroppo, R.G., & Nencioni, G. (2023a). 5G-MEC testbeds for V2X applications. *Future Internet*, 15(5), 175.

Wadatkar, P.V., Garroppo, R.G., & Nencioni, G. (2023b, October 2–6). *MigraMEC: Hybrid testbed for MEC app migration* [Paper presentation]. Annual International Conference on Mobile Computing and Networking (MobiCom), Madrid, Spain.

Wadatkar, P.V., Garroppo, R.G., Nencioni, G., & Volpi, M. (2024). Joint multi-objective MEH selection and traffic path computation in 5G-MEC systems. *Computer Networks*, 240.

World Economic Forum. (2020). *The impact of 5G: Creating new value across industries and society.* White Paper. https://www3.weforum.org/docs/WEF_The_Impact_of_5G_Report.pdf

Yago. (n.d.). *YAGO/Testarena Forus.* https://yago.no/hjem

Yaqoob, I., Hashem, I.A.T., Mehmood, Y., Gani, A., Mokhtar, S., & Guizani, S. (2017). Enabling communication technologies for smart cities. *IEEE Communications Magazine*, 55(1), 112–120.

Yigitcanlar, T., Han, H., Kamruzzaman, Md., Ioppolo, G., & Sabatini-Marques, J. (2019). The making of smart cities: Are Songdo, Masdar, Amsterdam, San Francisco and Brisbane the best we could build? *Land Use Policy*, 88, 104187.

Yigitcanlar, T., Mehmood, R., & Corchado, J.M. (2021). Green artificial intelligence: Towards an efficient, sustainable and equitable technology for smart cities and futures. *Sustainability*, 13(16), 8952.

12 Data accessibility for researchers in smart cities

A literature review and case study about access to consumer energy data in Norway

Helleik Rosenvinge Syse, Chandra Prakash Paneru, and Harald Nils Røstvik

Introduction

Never in human history has more data been available, and today over 50 billion devices collect data around the world (Sikimić et al., 2020). This data collection has led to an unprecedented amount of available data that can help researchers understand and mitigate pressing global challenges like climate change by providing more information about where energy-related emissions occur.

Reducing greenhouse gas emissions is critical to tackling climate change, and around 73% of global greenhouse gas emissions are linked to energy production (World Resources Institute, 2020). Therefore, decarbonizing energy production and using energy more efficiently are the most effective ways to combat climate change. However, to address the challenge of using energy more efficiently, data are needed to understand what happens at the consumer level (Jamasb & Meier, 2011).

Access to high-quality data is crucial not only for researchers but also for companies. However, companies do not necessarily benefit from sharing data outside the company. Often, it is the contrary, and some of the world's most valuable companies depend on data collection to retain and increase their value. For example, companies can use data to learn about an individual's behaviour, which can be used for marketing or increasing sales on their platforms. Also, evidence suggests that an increasing number of companies see the value of collecting data from their users to increase profits (Matthews, 2019).

Early papers on data collection and smart cities, like 'The Vision of a Smart City' (Hall et al., 2000), paint a bright future for the smart city. The smart city in this vision has technology at its core. It uses data to improve cities and facilitate collaboration between government, city managers, businesses, academia, and the research community. However, some argue that extensive data collection coupled with corporate interests is taking society in the opposite direction (Birch et al., 2021; Zuboff, 2018). See also Lorentzen and Langhelle, chapter 4, and Müller-Eie, chapter 7, which delve into this subject.

This chapter aims to identify barriers and propose solutions for researchers' access to consumer-level energy data. We carried out a case study to further

DOI: 10.4324/9781003498650-16

investigate the process of acquiring consumer-level energy data. This was done by contacting energy companies and searching for anonymized consumer-level energy data. A better understanding of the landscape of available energy data is crucial for research progress in this field. We want to present a concise review of the field and suggestions for improving energy data sharing across sectors.

Materials and methods

Research question: what are the potential barriers for researchers wanting to access consumer-level energy data?

To answer this research question, the study has two parts. First, a literature review is undertaken based on articles, books, institutional reports, and webpages that discuss the challenges researchers face in accessing data, specifically focusing on consumer-level energy data. Second, a case from the Norwegian context is studied, based on the authors' experience of trying to access data from five key players in the energy market for an individual research project focusing on the impact of fluctuating electricity prices on energy-saving behaviour. The data requests and meeting that make the basis for the case study took place between 2022 and 2023.

This chapter defines *consumer-level energy data* as 'electricity data measured at an individual household level with hourly resolution or better'. This type of data was collected from all households in Norway after the installation of smart meters became a requirement from 2015 onwards (Norwegian Water Resources and Energy Directorate, 2021). However, these data are only accessible to the individual electricity customer and the electricity providers, and, in some cases the electricity provider shares the data with third parties.

Literature search

To select the sources for the literature review, a screening process was implemented to ensure the inclusion of relevant materials while excluding those that did not significantly contribute to our research objectives. The screening process and keyword-based search were carried out according to the methodology outlined in Booth et al. (2012). For the grey literature search we employed methods from Adams et al. (2016). The following steps were taken to establish relevance and determine inclusion/exclusion:

1 Keyword-based search: we initiated our search process by querying Google Scholar. We employed various combinations of keywords to identify potential sources related to our research topic. These keywords included terms such as 'smart energy', 'limited access', 'open energy data', 'barriers', 'behaviour data', 'energy research', 'energy data', 'household energy', and 'energy policy'.
2 Initial screening: we screened the sources based on their titles and abstracts after the initial search. This initial screening helped us identify relevant materials aligned with our research focus.

3   Inclusion/exclusion criteria: we established a set of inclusion/exclusion criteria to refine our selection further. This was done through reading and systematizing the relevant sources based on the selected criteria:

- Published within the last 15 years.
- Relevant to the topics of household energy, data collection barriers, and privacy policy concerns.
- Presented content that was substantive and valuable to the discussion of challenges faced by researchers in collecting consumer-level energy data.

Sources that did not meet these criteria were excluded from our review.

4   Broadening the search: we wanted to include researchers' personal experiences by collecting behaviour-related data from grey literature. After completing the literature search, we found that few research articles described the obstacles and challenges researchers had when obtaining the energy data. Therefore, we decided to conduct a similar search in the grey literature using the Google search engine to glean insight into the personal experiences of researchers attempting to collect behaviour and energy data.

Case study

The process and experiences of the authors contacting and meeting with organizations with access to energy data are presented as the case study. The authors reached out to six companies from the Norwegian electricity market with varying sizes, ranging from start-up regional to national-level energy companies, both public and private, between 2022 and 2023. A standard data request was made to all the companies to access the empirical research data. The requests were made via one or more of the following channels: emails, physical meetings, and digital meetings. Within the companies we spoke to people with various roles. Sometimes, our networks were used to set up meetings with company representatives.

Literature review

What is the role of consumer-level energy data in smart city research?

Consumer energy data play a critical role in advancing smart city research, yet, as noted by Webborn et al. (2019), the availability of high-quality consumer energy data remains limited, restricting empirical studies. Despite these challenges, several behavioural intervention programs have utilized smart meters to encourage energy-saving behaviours. Smart meters, which represent an evolution of traditional energy meters, collect detailed information about end users' energy consumption directly from their devices (Zheng et al., 2013). Unlike conventional meters, smart meters can monitor real-time energy usage and facilitate two-way communication between the consumer and the energy system. This technology not only allows consumers to receive more accurate billing but also provides researchers with reliable data, thereby opening new opportunities for empirical research (Webborn et al., 2019).

The roll-out of smart meters changed how energy data are collected and increased the frequency of the feedback reaching the users. Before the smart meter, household energy consumption data were collected by reading the meter manually, and the data were then organized and sent to consumers with a bill (Klass & Wilson, 2016). European households used to get their electricity consumption data once a year or, at best, once a month (Leire et al., 2019).

The energy behavioural studies of the time before smart meters were primarily based on low-resolution annual or monthly billing data that did not account for the time-based variation resulting from various behavioural determinants (Kavousian et al., 2013). Empirical energy studies primarily relied on the anonymous data set of estimated energy consumption or fitted monitoring devices on individual buildings to gather consumption data (Webborn et al., 2019). In Norway, before the roll-out of smart meters, analogue meters were used to record the amount of electricity consumed (Livgard, 2010). Customers reported their analogue meter readings to energy providers. With the proliferation of smart meters, the readings became more accurate, the feedback frequency increased, and customers could monitor their consumption continuously (Livgard, 2010).

Further, advancements in sensor technology have led to increased use of smartphones and smart devices in the energy sector, including sensors, thermostats, and energy applications (Zhou et al., 2016). The penetration of these digital technologies has changed the landscape of the energy sector and enabled the collection of large amounts of energy production and consumption data. Globally, several energy providers have started developing smartphone-based energy apps that give customers insights into energy usage habits, historical energy consumption data, consumption in similar households, and real-time energy prices (Paneru & Tarigan, 2023).

With more data collected and stored, it has become possible to conduct analytics to discover electricity consumption patterns in smart cities (Pérez-Chacón et al., 2018). Although the power system has functioned historically without extensive data analytics, the increasing integration of variable renewable energy sources necessitates a more accurate and granular understanding of electricity consumption patterns. This data-driven approach is crucial for grid stability, cost-efficiency, and maximizing the potential of renewable energy. By identifying consumption patterns, we can optimize energy production, storage, and distribution, thereby enhancing the overall performance and resilience of the power system. Therefore, researchers use big data to forecast energy consumption and solar photovoltaic electricity production (de Freitas Viscondi & Alves-Souza, 2019). Big data do not necessarily imply cause-and-effect relationships but may reveal information about the context in which behaviour occurs (Tiefenbeck, 2017). According to Hong et al. (2017) and O'Brien et al. (2020), data collection is one of the biggest challenges of conducting behaviour research on human energy behaviour in buildings and cities.

Status of access to energy data for researchers

The literature reviewed describes several sources of household energy data useful for research. In a few instances, the full data sets are publicly available (Dong et al., 2022; Pullinger et al., 2021). However, generally, the data are not available publicly. We have chosen to group the sources of data into three broad categories: 'public data', 'data with limited access', and the largest category, 'private data'. We can use an iceberg as a metaphor to categorize these types of data. For real icebergs, 90% is below the waterline, while only 10% is above.

In this metaphor, the visible part of the iceberg are public data. Public data are easily available but often scarce and limited in usefulness, at least for household-level energy research. However, a vast expanse of data lies below the surface and is harder for researchers to reach. Data with limited access and private data are hidden beneath the surface. This iceberg figure of data availability is illustrated in Figure 12.1.

Confidential energy data accessibility, even for academic research, has historically been constrained by various factors, including privatization. In many countries, utility companies hold proprietary control over energy data, making

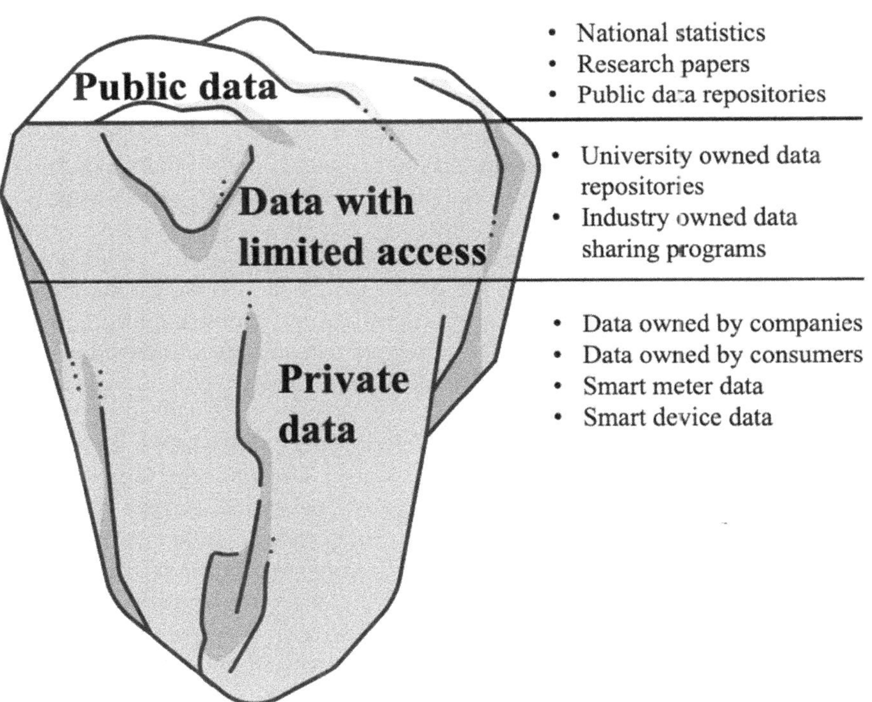

Figure 12.1 The 'iceberg' of energy data. Public data are at the top and most available. Below are data from sharing programs. At the bottom are 'private data', the most abundant form of data, which are also hardest to get access to.

researchers reliant on legislative changes or the goodwill of these companies for access (Webborn & Oreszczyn, 2019). Generally, energy companies may be hesitant to share energy data or behavioural insights with third parties as they fear that sharing the data might harm their competitive advantage or violate privacy regulations such as the General Data Protection Regulation (GDPR; Klass & Wilson, 2016). Specifically, they may be unwilling to share the data with researchers because they are aware that researchers scrutinize the data and make the results public through their publications. These conflicting concerns and interests regarding energy data between energy companies and researchers may be one of the reasons why it is difficult for researchers to obtain the data.

In the case of the United States, this lack of easy access to energy consumption data is a significant barrier to exploiting the smart grid's full potential (Klass & Wilson, 2016). The difficulty in acquiring data impedes various energy management programs, as energy companies control customer data, which are often unavailable for research. If researchers could access these data, their findings could inform evidence-based policy recommendations that would benefit both consumers and the industry.

Furthermore, although the rise of smart digital technology, such as smartphone-based energy apps, holds the potential to empower energy consumers in monitoring and controlling their energy use and promoting energy-efficient behaviours, independent researchers and policymakers face challenges in accessing and understanding data from such energy apps (Klass & Wilson, 2016). This limits the ability to study the apps' influence on consumer behaviour and their potential to promote energy efficiency.

The commercialization of energy data presents additional challenges. In the United Kingdom, a survey by Capgemini found that 84% of energy companies view data as a new revenue source, with many actively working to monetize consumer data (Husseini, 2019). This shift towards data commercialization, coupled with regulatory restrictions such as GDPR, makes it increasingly difficult for researchers to access domestic energy data without navigating complex legal and technical hurdles (O'Toole, 2021). For small research groups constrained by time, scope, and resource limitations, overcoming these obstacles can be particularly daunting (Webborn & Oreszczyn, 2019).

The situation is further complicated in United States cities, where several factors, including outdated data collection software, budget shortfalls, staff capacity limitations, and changes in political leadership, have been reported as barriers to accessing energy-related data (Aznar et al., 2015). In a comprehensive report by Salim et al. (2020), the authors investigated how data can be acquired to conduct research in this domain. According to these authors, one of the four major challenges when working with urban-scale behaviour data is '[the lack of] *a business model to motivate and enable creation and sharing of the data* [emphasis added]' (Salim et al., 2020, p. 12).

Despite limited progress with the proliferation of smart meters and advanced metering infrastructure, as well as increased data-sharing initiatives by utilities and regulators, challenges remain in obtaining granular energy consumption data for research purposes, with privacy concerns and fragmented data ownership being the main barriers (Yuan et al., 2024). Addressing these challenges will require a combination of legislative action, changes in data-sharing practices, and greater collaboration between energy companies and the research community. See also Huang et al., chapter 3 where institutional structures that promote cross-sector collaboration are suggested.

Case study

Efforts to access energy data: experience from Norway

We aimed to access anonymized household-level electricity consumption data from different Norwegian electricity companies offering consumers smart energy apps. The reason was that we intended to conduct empirical research on the impact of smart energy apps as smart digital solutions for Norwegian households' energy behaviour and energy-saving practices. Our approach involved several energy companies, each with its distinct portfolio in terms of ownership structure and market position. It included a mix of private, public, and publicly listed companies, ranging from regional start-ups to established national-level entities in the Norwegian electricity market. Through a series of formal data requests and collaboration efforts, we communicated our research objectives and data needs, resulting in various responses and challenges. Our primary communication was via email, which enabled us to communicate our research intentions and data requests simultaneously to key decision-makers at several energy companies. We also held digital meetings with representatives from three companies and personal interviews with employees at one company to clarify our requirements further. We sent them personalized emails about the purpose of our research and the types of data we were interested in. Follow-up emails were also sent to keep them informed about our requests. Table 12.1 shows the format of the data queries sent to each company, asking whether they were willing to share their data. Table 12.2 summarizes the responses received from the companies.

Table 12.1 Format of the request sent to energy companies asking about the possibility of data sharing for energy behavioural research

Data type	Active users
User demographic characteristics	1. Anonymous id
	2. Age
	3. Sex
	4. House/apartment type (size)
	5. City
Data duration	Last 12 months
Sample size (users)	500 continuous clients
Consumption, kWh (hourly basis)	User-wise
App engagement (daily basis)	No. of logins (user-wise)
Extra devices (e.g., smart home assistant, smart electric vehicle charger, solar panel)	List of devices (user-wise)

Table 12.2 Summary of the type of company and their response to the data sharing requests

Type of company	No. and type of representative(s)	Approximate time spent on meetings, communications, etc.	Summary of response
Private	2 (Customer service, media responsible)	Number of emails sent: 11 Time in meetings: 0.5 hour	Sharing user-specific data, even anonymously, with external parties is restricted by their terms and conditions. 'The legal department has stopped working on the data request due to GDPR concerns'.
Public	4 (Senior vice president, system developer, business developer, developer)	Number of emails sent: 6 Time in meetings: 2 hours	No idea where such data are stored; sent in circle between company representatives.
Publicly listed private	3 (Business developer, head of IT, coordinator)	Number of emails sent: 5 Time in meetings: 4.5 hours	At first very positive about sharing data but as time passed became more reluctant because sharing would increase the current workload.
Public	1 (CEO)	Number of emails sent: 2 Time in meetings: 0 hours	Effort to collect and anonymize the data is enormous, and they simply lack resources to help.
Private	2 (CEO and CTO)	Number of emails sent: 23 Time in meetings: 14 hours Preparing documents: 15 hours	Expressed interest in collaboration and was eager for a partnership with data sharing. Draft for research proposal to get external funding was made to cover costs. However, the collaboration was stopped due to economic hardships requiring different priorities from the company.

We, as researchers, have spent significant time, as shown in Table 12.2, trying to obtain household-level energy data. Currently we have no data that can be used to address the research challenges we first wanted to address. However, the meetings, communications, and follow-ups provided valuable insights into the motivations, challenges, and general process of obtaining this type of data from companies. This correspondence, in addition to the findings from the literature review, provides the basis for the following discussion.

Discussion

Background for study, sample size, and relevance

The case study has a relatively low number of studied cases (five electricity companies in total). Also, it is worth mentioning that the case studies started as a genuine effort to get hold of household-level energy data that the researchers would use for research projects, extending the scope of this chapter. However, as time passed and our efforts to get hold of this type of data continued to provide no results, the idea of writing this chapter on our failed efforts to obtain data emerged. Therefore, we took notes, kept an archive of our correspondence with the companies, and logged activities and the time spent. It was also crucial to anonymize the companies so the results presented could not be linked back to a specific company. The discussion and conclusions are based on both the literature review and the case study of contacting the companies. As the sample size is small, and the case study was not conducted as a study with formal interviews, no conclusions are drawn solely from the basis of the case study. However, the case study is used as an additional data point to relate the findings from the literature review to the Norwegian context and as a point of discussion concerning the findings from the relevant literature.

Public versus private sector views on sharing energy data

We received mixed responses in our attempts to reach out to various entities with access to household energy data. Our initial hypothesis was that there are differences in the attitudes and willingness to share energy data between the public and private sectors. We initially assumed that obtaining data from public entities would be easier than obtaining data from private companies, given the greater competition among private firms and their potential reluctance to disclose data that could confer a competitive advantage.

However, public energy companies seem to encounter challenges due to their complex data management structures, including a lack of technical knowledge and human resources to perform the tasks of exporting, processing, and sharing data or simply reluctance to increase workload. With private companies, GDPR issues, commercial interests, and resource constraints surfaced. One private company cited third-party data sharing restrictions as the constraint to sharing data, while others mentioned that anonymous data sharing was not an issue. These concerns were also reflected in the literature review.

Among the private companies, a common concern was whether there would be any benefit to sharing data. For instance, one of the companies was interested in using the potential collaboration to promote themselves among customers and create a positive impression. They highlighted the future possibility of creating an energy data set that they could share with universities and others to work on solutions that could benefit them further down the line. However, this positive development seems far-fetched based on their priorities and workload statements. They were also concerned about the extra workload involved in the process of data exporting and sharing and stated that they could not bear the additional workload now.

Although our findings did not directly support our assumption that the private sector is more hesitant to share data for competitive reasons, it is nevertheless a plausible explanation that also has some backing in the literature and should be investigated further. Generally, it is quite clear that energy research faces great difficulties in accessing data at the consumer level. These responses emphasise the need for an in-depth discussion of data-sharing practices in the energy sector.

The privacy policy paradox

In understanding the issue of researchers' access to energy data, we encountered a phenomenon we chose to name 'the privacy policy paradox' (PPP). This is not to be confused with the term 'the privacy paradox', which is the finding that most users are concerned about privacy online but undertake few actions to protect their data (Barth & De Jong, 2017).

The PPP, on the other hand, is the finding that recent policy laws like GDPR, whose primary aim is to *'enhance individuals' control and right over their personal data and to simplify the regulatory environment for international business* [emphasis added]' (European Union, 2015), in some instances do the opposite and, in addition, make it harder for researchers to access data. To elaborate further, because companies are afraid of breaching the regulations of, say, GDPR, they end up using regulations as an excuse not to share any data instead of finding work-arounds. This could lead to less publicly available data, which in turn could make it harder for researchers and start-ups that want to challenge the status of existing players because access to data is needed to build a working product or conduct research. For this reason, the PPP could also diminish individuals' control and rights over their personal data because one ends up in a situation close to a monopoly where access to personal data is restricted to a few big players.

This is not to say that privacy policies like GDPR in themselves are negative for individuals' privacy – quite the contrary. In our understanding, PPP usually boils down to misunderstandings, where companies claim that they cannot share any data because they are afraid of being fined for not complying with the GDPR. However, companies can share data if the data are fully anonymized or if the user opts to share it with researchers or other third parties. The big challenge is establishing easy processes for anonymous data sharing and incentivizing businesses to share their data.

Proposed solutions

According to Pfenninger et al. (2017), energy research data sharing has lagged. Understandably, the difficulty in accessing energy data is partly linked to underlying ethical and security concerns, unintended exposure, extra workload, and institutional or personal unwillingness (Pfenninger et al., 2017). However, to support data-driven science and evidence-based policy recommendations, it is necessary to find solutions to solve the data access challenges and the underlying hurdles for independent researchers to obtain the energy data. Policymakers could support companies to create a common platform for sharing data sets, potentially easing researchers' data collection difficulties. This development could benefit energy companies, academia, and the public sector.

Some methods have already been explored and implemented in various countries to relieve researchers' burden of energy data collection. In the UK, to facilitate researchers' easy access to and use of the smart meter data, a Smart Energy Research Lab (SERL) was conceptualized, which consisted of a consortium of six Great Britain universities and the Energy Saving Trust. The SERL lifted the heavy technical, legal, and data governance steps so that researchers do not have to jump hurdles to access the data. SERL operated as a central research data resource for the UK research community, and they recruited over 13,000 UK households who agreed to share data. Researchers can access the data if they are accredited by the Office for National Statistics and are part of any UK research institution (O'Toole, 2021).

There are also examples of programs that aim to crowdsource data, like the *Donate Your Data* program, administered by ecobee Inc. This program contains over 3 million days of data from more than 112,000 thermostats across North American homes. It consists of measurements recorded at 5-minute intervals for indoor and outdoor air temperatures, set points, indoor relative humidity, motion sensors, and HVAC (heating, ventilation, and air conditioning) equipment run times (Salim et al., 2020). However, concerns such as crowdsourced data being susceptible to self-selection bias must be critically accessed. Also, when the authors applied for access to this program, they did not receive a reply from the company.

Outside the energy sector, various initiatives have already been practised within healthcare for a better operational mechanism for data sharing with researchers (Hulsen, 2020). Although healthcare is more privacy and security sensitive, if it can establish an effective operating mechanism, then the best practices from healthcare and other sectors could be transferred to the energy sector, which will require comprehensive studies and policies from the concerned policymakers.

Conclusions

In this study, we explored the significant barriers researchers face in accessing high-resolution energy data, which is crucial for advancing energy efficiency research and promoting pro-environmental behaviours. Our findings highlight these issues: the absence of an established system for large-scale energy data sharing that ensures privacy, the growing reluctance of companies to share data due to its increasing value, and the additional burden on companies to anonymize data for research purposes. In addition, the lack of economic incentives for data sharing further complicates access to data.

These challenges limit the empirical validation of theories and the development of data-driven policy recommendations. Our discussion suggests that while some companies are open to data sharing and there are not necessarily any legal barriers to doing so, the system has been set up so that private companies own the data and have no incentives for sharing it with researchers. Some companies are positive about sharing data but cite time constraints on data export and anonymization as a barrier. Countries like the UK and the United States have recognized existing barriers and implemented some initiatives to ease them. However, many countries have not yet identified these barriers.

Moreover, the PPP can explain why accessing energy data is challenging for researchers. According to this explanation, privacy policies such as GDPR may restrict researchers' access to energy data due to a misinterpretation of the rules. If less data are made available to the public and researchers, this may paradoxically lead to individuals having less control over the data, as only a few companies can see the complete picture of what is happening in the energy system and can use the data to their advantage.

There is a need for further research to explore these issues in depth, as understanding energy consumption at the household level is critical to developing effective policies to reduce emissions. This study has highlighted existing challenges, but addressing these barriers is crucial for unlocking the full potential of energy data in mitigating global climate change.

Acknowledgements

Special thanks go to the Smart City Network at the University of Stavanger for their support with the PhD course BYG905, which sparked the idea for this chapter. Although our efforts to access data have not succeeded, we acknowledge the energy companies' willingness to collaborate and their contributions.

Funding

Helleik R. Syse is part of the Future Energy Hub Project funded by the Norwegian Research Council (Project No. 280458). Chandra Prakash Paneru is part of the Greencoin project that has received funding from the Applied Research – Cities for the Future: Services and Solutions programme (under Grant Agreement No. NOR/IdeaLab/GC/0003/2020–00).

Conflicts of interest

The authors declare no conflict of interest.

References

Adams, R., Smart, P., & Huff, A. (2016). Shades of grey: Guidelines for working with the grey literature in systematic reviews for management and organizational studies. *International Journal of Management Reviews*, 19(4), 432–454. doi:10.1111/ijmr.12102

Aznar, A., Day, M., Doris, E., Mathur, S., & Donohoo-Vallett, P. (2015). *City-level energy decision making: Data use in energy planning, implementation, and evaluation in U.S. cities. National Renewable Energy Laboratory.*

Barth, S., & De Jong, M.D. (2017). The privacy paradox – Investigating discrepancies between expressed privacy concerns and actual online behavior – A systematic literature review. *Telematics and Informatics*, 34(7), 1038–1058.

Birch, K., Cochrane, D., & Ward, C. (2021). Data as asset? The measurement, governance, and valuation of digital personal data by Big Tech. *Big Data & Society*, 8(1). doi:10.1177/20539517211017308

Booth, A., Papaioannou, D., & Sutton, A. (2012). *Systematic approaches to a successful literature review.* Sage.

de Freitas Viscondi, G., & Alves-Souza, S.N. (2019). A systematic literature review on big data for solar photovoltaic electricity generation forecasting. *Sustainable Energy Technologies and Assessments*, 31, 54–63. doi:10.1016/j.seta.2018.11.003

Dong, B., Liu, Y., Mu, W., Jiang, Z., Pandey, P., Hong, T., Olesen, B., Lawrence, T., O'Neil, Z., Andrews, C., Azar, E., Bandurski, K., Bardhan, R., Bavaresco, M., Berger, C., Burry, J., Carlucci, S., Chvatal, K., De Simone, M., … Zhou, X. (2022). A global building occupant behavior database. *Scientific Data*, 9(1), 369. doi:10.1038/s41597-022-01475-3

European Union. (2015). *EN Proposal for a regulation of the European Parliament and of the Council on the protection of individuals with regard to the processing of personal data and on the free movement of such data (General Data Protection Regulation)*. https://data.consilium.europa.eu/doc/document/ST-9565-2015-INIT/en/pdf

Hall, R.E., Bowerman, B., Braverman, J., Taylor, J., Todosow, H., & Von Wimmersperg, U. (2000). *The vision of a smart city* [Paper presentation]. 2nd International Life Extension Technology Workshop, Paris, France.

Hong, T., Yan, D., D'Oca, S., & Chen, C.-f. (2017). Ten questions concerning occupant behavior in buildings: The big picture. *Building and Environment*, 114, 518–530. doi:10.1016/j.buildenv.2016.12.006

Hulsen, T. (2020). Sharing is caring – Data sharing initiatives in healthcare. *International Journal of Environmental Research and Public Health*, 17(9), 3046.

Husseini, T. (2019). *As energy companies race to cash in on data, should we be worried?* Power Technology. https://www.power-technology.com/analysis/energy-companies-sharing-data/

Jamasb, T., & Meier, H. (2011). 13 Energy spending and vulnerable households. *The Future of Electricity Demand: Customers, Citizens and Loads*, 69, 318.

Kavousian, A., Rajagopal, R., & Fischer, M. (2013). Determinants of residential electricity consumption: Using smart meter data to examine the effect of climate, building characteristics, appliance stock, and occupants' behavior. *Energy*, 55, 184–194. doi:10.1016/j.energy.2013.03.086

Klass, A.B., & Wilson, E.J. (2016). Remaking energy: The critical role of energy consumption data. *California Law Review*, 104, 1095.

Leire, B., Jed, J.C., Andrea, K., Ana, M., & Johannes, R. (2019). Exploring the role of ICT on household behavioural energy efficiency to mitigate global warming. *Renewable and Sustainable Energy Reviews*, 103. doi:10.1016/j.rser.2019.01.004

Livgard, E.F. (2010, July 4–7). *Electricity customers' attitudes towards smart metering* [Paper presentation]. 2010 IEEE International Symposium on Industrial Electronics, Bari, Italy.

Matthews, D. (2019, October 23). Researchers concerned as tech giants choke off access to data. *Times Higher Education*. https://www.timeshighereducation.com/news/researchers-concerned-tech-giants-choke-access-data

Norwegian Water Resources and Energy Directorate. (2021). *Smarte strømmålere (AMS)*. https://www.nve.no/reguleringsmyndigheten/kunde/strom/stromkunde/smarte-strommalere-ams/

O'Brien, W., Wagner, A., Schweiker, M., Mahdavi, A., Day, J., Kjærgaard, M.B., Carlucci, S., Dong, B., Tahmasebi, F., Yan, D., Hong, T., Gunay, H.B., Nagy, Z., Miller, C., & Berger, C. (2020). Introducing IEA EBC annex 79: Key challenges and opportunities in the field of occupant-centric building design and operation. *Building and Environment*, 178, 106738. doi:10.1016/j.buildenv.2020.106738

O'Toole, J. (2021). *Unlocking the potential of smart meter data to accelerate smart local energy systems*. https://www.energyrev.org.uk/news-events/blogs/unlocking-the-potential-of-smart-meter-data-to-accelerate-smart-local-energy-systems/

Paneru, C.P., & Tarigan, A.K.M. (2023). Reviewing the impacts of smart energy applications on energy behaviours in Norwegian households. *Renewable and Sustainable Energy Reviews*, 183, 113511. doi:10.1016/j.rser.2023.113511

Pérez-Chacón, R., Luna-Romera, J.M., Troncoso, A., Martínez-Álvarez, F., & Riquelme, J.C. (2018). Big data analytics for discovering electricity consumption patterns in smart cities. *Energies*, 11(3), 683. https://www.mdpi.com/1996-1073/11/3/683

Pfenninger, S., DeCarolis, J., Hirth, L., Quoilin, S., & Staffell, I. (2017). The importance of open data and software: Is energy research lagging behind? *Energy Policy*, 101, 211–215. doi:10.1016/j.enpol.2016.11.046

Pullinger, M., Kilgour, J., Goddard, N., Berliner, N., Webb, L., Dzikovska, M., Lovell, H., Mann, J., Sutton, C., Webb, J., & Zhong, M. (2021). The IDEAL household energy dataset, electricity, gas, contextual sensor data and survey data for 255 UK homes. *Scientific Data*, 8(1), 146. doi:10.1038/s41597-021-00921-y

Salim, F.D., Dong, B., Ouf, M., Wang, Q., Pigliautile, I., Kang, X., Hong, T., Wu, W., Liu, Y., Rumi, S.K., Rahaman, M.S., An, J., Deng, H., Shao, W., Dziedzic, J., Sangogboye, F.C., Kjærgaard, M.B., Kong, M., Fabiani, C., … Yan, D. (2020). Modelling urban-scale occupant behaviour, mobility, and energy in buildings: A survey. *Building and Environment*, 183, 106964. doi:10.1016/j.buildenv.2020.106964

Sikimić, M., Amović, M., Vujović, V., Suknović, B., & Manjak, D. (2020, March 18–20). *An overview of wireless technologies for IoT network* [Paper presentation]. 2020 19th International Symposium INFOTEH-JAHORINA (INFOTEH), East Sarajevo, Bosnia and Herzegovina.

Tiefenbeck, V. (2017). Bring behaviour into the digital transformation. *Nature Energy*, 2 (6), 17085. doi:10.1038/nenergy.2017.85

Webborn, E., Elam, S., McKenna, E., & Oreszczyn, T. (2019). Utilising smart meter data for research and innovation in the UK. In *Proceedings of ECEEE 2019 Summer study on energy efficiency* (pp. 1387–1396). ECEEE.

Webborn, E., & Oreszczyn, T. (2019). Champion the energy data revolution. *Nature Energy*, 4(8), 624–626. doi:10.1038/s41560-019-0432-0

World Resources Institute. (2020). *Climate watch.*

Yuan, R., Pourmousavi, S.A., Soong, W.L., Black, A.J., Liisberg, J.A.R., & Lemos-Vinasco, J. (2024). Unleashing the benefits of smart grids by overcoming the challenges associated with low-resolution data. *Cell Reports Physical Science*, 5(2), 101830. doi:10.1016/j.xcrp.2024.101830

Zheng, J., Gao, D.W., & Lin, L. (2013, April 4–5). *Smart meters in smart grid: An overview* [Paper presentation]. IEEE Green Technologies Conference (GreenTech), Denver, CO, USA.

Zhou, K., Fu, C., & Yang, S. (2016). Big data driven smart energy management: From big data to big insights. *Renewable and Sustainable Energy Reviews*, 56, 215–225. doi:10.1016/j.rser.2015.11.050

Zuboff, S. (2018). *The age of surveillance capitalism: The fight for a human future at the new frontier of power.* Profile Books.

13 Barriers, motivators, and smart solutions for promoting commute cycling in Stavanger

Ayda Joudavi and Ari K.M. Tarigan

Introduction

Smart solutions have been a crucial part of creating sustainable and liveable cities. Over the past few decades, the concepts of smart solutions and sustainable cities have been blended as 'smart city'. *Smart city* often refers to an urban area using information and communication technologies (ICTs), business models, and solutions to increase operational efficiency, share information with the public, and improve the quality of services and citizen welfare (Brčić et al., 2018).

However, the definition of a smart city varies depending on the approach (see also Müller-Eie, chapter 7). Neoliberal models prioritize private-sector solutions, such as public–private partnerships and technology-driven services, focusing on profitability and efficiency (Mora et al., 2017). In contrast, public interest approaches emphasize equity, inclusivity, and the common good, aiming to leverage technology for the benefit of all citizens rather than maximizing economic returns (Mora et al., 2017).

Smart city solutions help planners to better understand transportation patterns and manage transportation needs by collecting relevant data through fixed and mobile sensors. They are often associated with sustainable transportation and citizens' well-being (Poslad et al., 2015). Smart mobility, as one of the main pillars of the *smart city* concept, has evolved considerably over the past few years. Smart mobility includes solutions driven by innovations through digital technology that help improve how people travel to become more effective and sustainable (Benevolo et al., 2016; Maldonado Silveira Alonso Munhoz et al., 2020; Paiva et al., 2021). Smart mobility may use digital technology solutions such as in-vehicle navigation tools, information-mobility signage (digital displays to communicate information), autonomous vehicles, demand-responsive transport, carpooling services, live tracking, biometric scanning, dynamic queuing management, e-parking, e-ticketing, and e-passes. Many of these solutions are widely used in urban environments today (Šurdonja et al., 2020).

Active mobility, such as cycling, is often considered a reliable alternative for reducing car use and is an inseparable part of any sustainable mobility plan (Lättman & Otsuka, 2024). Cycling is an option for many commuters (e.g., those who only must travel short distances) and is often associated with environmental and

DOI: 10.4324/9781003498650-17

health benefits (Cervero & Duncan, 2003). Cycling has become the backbone of urban transportation in many cities worldwide that aim to target climate neutrality. Cycling is also favourable for citizens' well-being as it is relatively low-cost and promotes a healthy lifestyle. However, there are still challenges and barriers to increase cycling (Semenescu & Coca, 2022).

Smart solutions that depend on digital technologies, such as electric bike sharing, gamification initiatives, and activity and health tracking, have recently been introduced in the cycling sector to attract more cyclists and are gaining increasing attention (L. Pan et al., 2022; Pasca et al., 2021). Among measures to promote cycling as beneficial to both climate neutrality and well-being, compared to other means of transport for everyday commuting (Teixeira et al., 2021), facilitating digital technologies as smart solutions for cycling appears promising (Krizek et al., 2009).

This theoretical study contributes to increasing insight into these interrelations. A literature review is used to explore the barriers and motivators underlying citizens' preference for cycling as a mode of commuting travel and to explore the use of smart solutions to promote cycling. Further, this study qualitatively and theoretically explores the smart and digital solutions that have been adopted in Stavanger Municipality to encourage commute cycling. The study addresses the question:

> Which smart solutions are utilized in Stavanger Municipality's smart initiatives to promote biking, and how may these initiatives be further developed to contribute to overcoming barriers to commute cycling?

The chapter offers critical reflections and discussions on evidence reported in Stavanger and challenges for the municipality in promoting cycling for the daily commute.

A subjective assessment approach

The study is based on a qualitative traditional literature review, inspired by M. L. Pan (2016) and Rozas and Klein (2010) and following the approach by Krizek et al. (2009). The review was based on scientific articles from the last 15 years. We employed two main online databases, Web of Science and Scopus, to screen the relevant articles. The search was limited to transportation-related publication sources with recognized credibility to limit the review's focus. We only included articles written in English and excluded grey literature, such as conference papers and technical reports. Further, this study provides a subjective assessment based on a review of Stavanger Municipality's related policy and technical documents and the authors' experiences as citizens and mobility researchers.

General motivation for and barriers to cycling

According to Rérat (2021), convenience, physical and mental well-being, a positive social image, and independence are among the key motivators for cycling. Several studies have examined the factors influencing convenience, such as the proximity of bike stations to workplaces and residential areas (Bachand-Marleau et al., 2012; Fishman et al., 2015; Hosford & Winters, 2018; Raux et al., 2017), being able to save travel time (Fishman, Washington, & Haworth, 2014; Reilly et al., 2020; Ricci, 2015), accessibility and ease of use of the bike and e-bike sharing schemes (Fishman et al., 2012; Wu et al., 2021), and the possibility for using a bike jointly with other modes of transport (Murphy & Usher, 2015).

Some studies have reflected on environmental and health benefits as strong reasons for choosing cycling over other modes of transport (Cerutti et al., 2019; Fishman, Washington, & Haworth, 2014; Hosford et al., 2018; L. Wang, 2018). Other studies have emphasized the advantages of biking for mental well-being and the joy of riding a bike (S. Y. Chen & Lu, 2016; Fishman, Washington, Haworth, & Mazzei, 2014; Heinen & Handy, 2012; Heinen et al., 2010; Rérat, 2019). The importance of social engagement and having a supportive social environment to sustain cycling behaviour has also been discussed (Z. Chen et al., 2020; Fishman et al., 2012; Teixeira et al. 2021; L. Wang, 2018). Additionally, the flexibility and freedom, including the ability to choose routes and travel times, make cycling an appealing alternative to a car (Gössling, 2013; Heinen et al., 2010; Mackett, 2014; Pucher et al., 2010; Rérat, 2019).

On the other hand, several studies have explored common barriers to cycling, including the lack of adequate cycling infrastructure (Buck & Buehler, 2012; Fishman et al., 2012; Noland et al., 2016), safety concerns (Fishman, 2016; Rérat, 2019; Ricci, 2015), long distances to destinations (Rérat, 2019), route characteristics such as hilliness and safety (Teixeira et al., 2021), unsuitable weather conditions, and the absence of facilities, such as showers and bike parking, at destinations (Fishman, Washington, Haworth, & Mazzei, 2014; Iwińska et al., 2018; Teixeira et al., 2021). Among these barriers, safety concerns are the most frequently cited in the literature.

In the context of smart cities, cycling often integrates digital components to enhance the cycling experience and promote sustainable urban mobility. These digital features apply technology to cycling, including route planning, safety, accessibility, and community interaction (Aderibigbe & Gumbo, 2024; Giffinger et al., 2007; Gonçalves et al., 2021; Huovila et al., 2019).

Types of ICT-based smart solutions in promoting cycling

The literature addresses six key 'smart solutions' provided by ICT that seem to promote cycling.

Real-time travel information

Many studies, such as Balakrishna et al. (2013), Dziekan and Kottenhoff (2007), and Kurauchi et al. (2019), have emphasized the importance of providing real-time travel information to people to change their travel behaviours; for example, people's decisions regarding the mode, route, and departure time choice (Poslad et al., 2015). Real-time travel information offers cyclists valuable data on routes, weather conditions, and safety, enhancing the ease and predictability of cycling as a commuting option (Ayad et al.,

2024; Böcker et al., 2015; Zhao et al., 2018). Real-time travel information exists on open-source platforms such as Google Maps, and it often connects with different platforms as a supporting infrastructure to improve the performance of the main services, such as bike sharing schemes, multimodal transport systems, and recreational activity platforms (e.g., Strava; Albuquerque et al., 2021). Overall, it is argued that real-time information is beneficial to help cyclists effectively reach their destinations.

Personalized travel planning

Personal travel planning involves providing tailored information and incentives while encouraging personal commitment and feedback (Müller-Eie et al., 2019). It is especially effective in shifting travel behaviour, helping reduce car trips in favour of increased cycling (Kramers et al., 2014; Müller-Eie et al., 2019). Mobile applications, such as those on smartphones, can serve as convenient platforms for delivering this information directly to individuals. Studies indicate that when cyclists are better informed about their travel choices, they are more inclined to choose cycling over other modes of transportation (Kramers et al., 2014). However, the impact of these systems on significantly reducing car use remains uncertain, as success largely depends on the availability of alternative transport options and the quality of the supporting infrastructure (Šurdonja et al., 2020).

Feedback and self-monitoring

Many studies (Festinger, 1954; Sunio & Schmöcker, 2017; Swan, 2012) have emphasized the importance of feedback in altering and sustaining behavioural change. For example, feedback allows individuals to monitor their behaviour (Buchanan et al., 2014) and review the impact of their travel choices on the environment in terms of carbon footprints and greenhouse gas emissions and for themselves in terms of monthly cost, health benefits, and time saved (Poslad et al., 2015; L. Wang, 2018).

A wide range of applications allow users to set personal targets like weight loss or CO_2 reduction and track their progress towards the goal (Jakicic et al., 2016; Wei et al., 2021). In addition, these apps often provide them with useful tips on achieving their goals. An example is a smartphone application called 'Capture', which allows users to register their travel data and track their carbon footprint; the app also delivers personalized suggestions and tips for reducing emissions or offers to offset one's emissions by engaging in verified carbon offsetting projects around the world (Bearne, 2021). These tools provide immediate feedback on personal and environmental benefits, helping to sustain behaviour change. Nevertheless, their long-term effectiveness in maintaining increased cycling rates is questionable, as intrinsic motivation may wane once the novelty of the technology fades (Buchanan et al., 2014).

Social networks

Different studies (Batterbury, 2003; Binsted & Hutchins, 2012; Bonneau & Preibusch, 2010; Cho et al., 2011) have investigated the impact of peer influence on travel behaviour. It has been argued that social networks can enhance and facilitate travel behaviour change and connect people with similar travel patterns, interests, or environmental concerns (Thogersen, 2007). Using social networks, people can share their travel experiences and information regarding mode choice, route choice, travel cost, and safety concerns. Social networks also allow users to share their performance statistics with others, triggering competition and encouraging others to choose more active travel. Peer influence can amplify the willingness to change travel behaviour and promote behavioural change (Poslad et al., 2015).

The role of social networks and peer influence in promoting cycling has also been highlighted, with platforms like Strava allowing cyclists to share their achievements and encourage others (Lee & Sener, 2021). While these platforms promote cycling through competition and social engagement, their impact is often limited to those already inclined toward cycling, making it difficult to reach new users who may not be as motivated by social comparison (Thogersen, 2007).

Rewards and points

Reward systems, including point-based schemes and competition, gamification, and monetary incentives, have been implemented in travel studies to facilitate cycling (Ball et al., 2017; Kazhamiakin et al., 2015; W. Wang et al., 2022; Yang et al., 2010) and have been shown to be successful (Ben-Elia & Ettema, 2009; Bliemer & van Amelsfort, 2008). However, they come with an economic cost that makes sustaining the behavioural change less feasible (Poslad et al., 2015), and they often do not result in lasting behavioural change once the incentives are removed, reflecting the limits of extrinsic motivators (Ben-Elia & Ettema, 2009).

Digital social market

A digital social market is a smartphone application that enables individuals to exchange information and interact with each other (Manca et al., 2022). These platforms can also be used to promote pro-environmental behaviours (Cellina et al., 2020; White & Marchet, 2021). However, they include equity concerns regarding potential exclusion of people, privacy concerns resulting from collecting personal data, and uncertainty regarding the long-term viability of reward programs. Further research and assessment are required regarding the impact of a digital social market on promoting travel behaviour change towards cycling (Manca et al., 2022).

Smart solutions for promoting biking in Stavanger, Norway

Stavanger and the surrounding region, Nord Jæren, has a population of about 250,000 inhabitants (Brinkhoff, 2024). In Stavanger, the share of bicycle use is about 7% of the total transport mode choice (Nasjonal reisevaneundersøkelse, 2021). This share is rather low, despite appropriate cycling infrastructure, such as Norway's most expensive bicycle road (estimated cost ca. 175 million euro; Taranger et al., 2023), red bike lanes (the right-most lane on most of the main streets in the city; Stavanger Municipality, 2024a), and red bike streets (where 80% of the street is dedicated to cyclists; Stavanger Municipality, 2024b). To increase cycling, the city has adopted various 'smart' initiatives, including smart solutions:

Stavanger municipality's smart initiatives to promote biking

Stavanger has adopted a few smart mobility initiatives, mainly through the local mobility (bus, ferries, train) provider, Kolumbus (Kolumbus, 2022a). Kolumbus has integrated innovative digital solutions like real-time travel information, bike sharing, and combined ticketing solutions into its existing public transport services (Kolumbus, 2022b) to support Stavanger's goal of becoming a low-carbon-emission society by 2050, which aligns with Norway's national environmental goals (Skoglund, 2019; Stavanger Municipality, 2018). Some of the smart initiatives which Stavanger municipality has adopted for this aim are described below.

Real-time information

Kolumbus has developed an app (called 'Sanntid' in Norwegian) which offers real-time information about buses, trains, and city bikes to facilitate journey planning. This real-time system gets its data from different sensors installed on buses or bikes and precisely calculates how long it takes for a bus to reach a stop, how many city bikes are available in the station, and how much charge they have (Kolumbus, 2019). Although Sanntid is not directly associated with cycling, this real-time information indirectly helps commuters to avoid unnecessary waiting time by combining cycling and public transport during commuting journeys. Kolumbus also provides information screens that share real-time information about transport options, departure and arrival times, and delays in public places (Kolumbus, 2019).

City bikes (Bysykkel)

Stavanger had more than 750 electric city bikes by August 2021 (Pritchard & Lovelace, 2022), used in a Stavanger region–wide bike-sharing project to make public transport easier, even if there is no bus or train stop in the immediate vicinity. Bikes are available in many places and can be picked up at one place and delivered to another (Kolumbus, 2021a).

The bikes are equipped with built-in GPS and a digital locking system to check their availability in nearby places, to track and locate bikes not delivered to charging stations, and to notify when a bike runs on a low battery (Kolumbus, 2021a).

Renting bikes takes place on the same app used for buying public transport tickets. It costs 0.8 euros to unlock and 0.085 euros per minute to ride the bike (Kolumbus, 2022b); a 15-minute bike ride is free if one has an active public transport ticket. There are 145 charging stations installed throughout the city (Kolumbus, 2022b), and the municipality plans to make them also available for the charging of private e-bikes (Pritchard & Lovelace, 2022).

Combined ticketing

Kolumbus introduced a combined low-cost mobility ticket, including free access to electric city bikes, to use to and from trains, buses, and ferries (Kolumbus, 2022b, 2024a).

A 2021 Kolumbus survey about user (n = 1,259) satisfaction and use of the city bike scheme revealed that only one-fifth of city bike users pay directly for using this service, while around 42% use a regular ticket that includes a 15-minute free ride, and 33% use a HjemJobbHjem (Home–Work–Home) discounted ticket that includes a 60-minute free ride on a city bike. These findings underscore the positive impact of integrated ticketing solutions on promoting city bikes as an effective and convenient 'last mile' travel option in conjunction with public transportation (Pritchard & Lovelace, 2022). However, more than 63% of the respondents to Kolumbus's satisfaction survey had experienced barriers to the city bike use because of empty stations or bicycles in stations that cannot be rented (due to low charging levels or technical issues; Pritchard & Lovelace, 2022).

Home-Work-Home (HjemJobbHjem)

HjemJobbHjem (HJH) is a mobility initiative funded by the government (Government.no, 2021) and is offered to companies located in the urban region of Nord-Jæren. Employees from affiliated companies can buy monthly passes for a mobility package that enables them to use trains, buses, and city bikes for an even lower price than the combined ticket and for a full hour of a free ride on electric city bikes. By 2022, more than 34% of the workforce in Nord-Jæren were working for an organization affiliated with the HJH scheme (Kolumbus, 2022b). However, the price difference between monthly passes and HJH tickets (about 10 euro; HjemJobbHjem, 2024a) may not be significant enough to influence individuals' decisions to leave the car at home. Moreover, HJH tickets lack some of the benefits of regular tickets, such as free boarding for one adult companion and up to three children after 5 p.m. This feature is particularly appealing for families with small children, and its absence on HJH tickets could negatively impact individuals' decisions to use them. However, introducing the 60-minute free ride city bikes for HJH members has increased the use of city bikes, and in 2021 33% of city bike users were HJH ticket holders (Pritchard & Lovelace, 2022).

Cycling campaign (Sykkelløftet/Stiim)

Sykkelløftet (a collaboration between Rogaland County Council, Stavanger and Sandnes municipalities, the Norwegian Public Roads Administration, and business associations) was a campaign program developed by Netpower (Netpower, 2019), primarily as a web-based solution to encourage employees in the region to cycle to work. Participants could earn points by registering on the website and keeping records of how they got to and from work and win prizes at the end of each campaign (Eiterjord et al., 2011). A later incorporation of this initiative into a mobile application called 'stiim' allowed users from the same company to create groups, make teams, and earn points and scores for travelling to work using public transport or bike (Kolumbus, 2024b). However, the long-term effectiveness of such a gamification approach to change travel behaviour is unclear, especially its attraction to cycling newbies and people with low willingness to cycle (Kolumbus, 2021b).

Health project (Helseprosjekt)

Various HJH-affiliated companies joined a health project (Helseprosjektet) to encourage active mobility, targeting employees motivated to embrace a healthier life style and to use walking and cycling for their work trips (HjemJobbHjem, 2024b). This project offered assessment of participants' health conditions before and after the project period (usually 12 weeks), free or low-priced use of public transport or e-bikes, and social networking on Facebook to share their activity statistics and how they experienced being involved in this project (HjemJobbHjem, 2021). According to a post-intervention survey, 53% of the health project participants stated that they continued biking to work after the project period (HjemJobbHjem, 2019). However, a 12-week period is relatively brief to evaluate the long-term impact of such initiatives on travel behaviour. Furthermore, since participants in the health project join voluntarily, the initiative's impact on those less inclined to engage in cycling remains unexplored.

Table 13.1 summarizes which key ICT-based smart solutions are utilized in Stavanger Municipality's smart initiatives (described above) to promote biking.

This overview (Table 13.1) shows that providing real-time travel information and personalized travel planning are the functions which are utilized in most of the smart mobility initiatives in Stavanger. The Stavanger cycling campaign and the health project provide feedback and self-monitoring, social networks, and rewards and point functions, while the cycling campaign solution can provide an

Table 13.1 A review of smart solutions for promoting cycling in Stavanger

Implemented smart solutions for cycling in Stavanger	*Functions (types) of ICT-based smart solutions in promoting cycling*					
	Real-time travel information	*Personalized travel planning*	*Feedback and self-monitoring*	*Social networks*	*Rewards and points*	*Digital social market*
Real-time Information, Sanntid	X	X				
City bikes	X	X				
Combined ticketing	X	X				
HJH	X	X				
Cycling campaign			X	X	X	X
Health project			X	X	X	

additional function, the digital social market. None of the initiatives utilized the entire potential of available ICT-based smart solutions.

Concluding discussion

This study has assessed which different smart mobility solutions Stavanger Municipality has utilized in its initiatives to promote biking. The initial review highlighted the growing international recognition of ICT-bases smart solutions to boost the motivators for cycling or address the barriers to cycling (Aderibigbe & Gumbo, 2024; Giffinger et al., 2007; Gonçalves et al., 2021; Huovila et al., 2019; Šurdonja et al., 2020; L. Pan et al., 2022; Teixeira et al., 2021) and their promising influences on commute cycling (Krizek et al., 2009).

This concluding discussion critically reflects on how Stavanger Municipality may further develop its initiatives to contribute to overcoming barriers to commute cycling.

The assessed initiatives demonstrate Stavanger's commitment and effort to promoting sustainable alternatives for daily commuting, particularly cycling. However, in Stavanger, the offered smart solutions (HJH, real-time information [Sanntid], combined ticketing, and city bikes) are all provided by the local mobility operator Kolumbus. This entails complex challenges, as such a mobility company may prioritize other goals, such as an increase in public transport, rather than promoting cycling. Kolumbus seems to focus on providing real-time information and personalized travel planning rather than providing the other types of ICT-based functions.

In this connection, a relatively new Kolumbus payment solution, with fares calculated based on the actual distance travelled rather than traditional zone tickets (Kolumbus, 2022a), challenges privacy concerns, rather than promoting cycling. This is because the new system requires public transport users to share their location with Kolumbus so the company can track their positions and journeys. While this approach may offer financial benefits to users and may improve city bike services, it raises concerns about both privacy and the data being collected (Eckhoff & Wagner, 2017; Ismagilova et al., 2022).

City bikes and the health project are two of the Stavanger initiatives that could benefit from the integration of further ITC-based smart solution to improve their targeting to promote bicycle use. Cyclists may be further motivated to use city bikes through integrated possibilities to record feedback and self-monitoring about how far and how often they cycle regularly.

Such information may be a crucial contribution to health monitoring and personal information amongst cyclists. City bikes could also create social networks amongst the cyclist community, which might be highly relevant as a motivator and may be an effective feature for digital social marketing. Lastly, city bikes could offer a reward and point system to incentivize non-regular cyclists to be regular cyclists. Similarly, in the health project, the cyclists could collect feedback related to their health improvement after the cycling experience, create social networks, and gain rewards and points, which later may trigger them to participate more regularly in cycling and other health-promoting activities, even if long-term effectiveness is still uncertain (Buchanan et al., 2014).

Despite its positive reception, the city bike system is still not extensive enough to be a reliable option for daily commuting (Pritchard & Lovelace, 2022). To enhance the user experience and make city bikes a viable option for daily commutes, the number of active bikes, as well as the number and distribution of city bike stations, should be increased.

Providing real-time travel information across the public transport fleet, including buses, trains, and city bikes, is another feature that aims to enhance the commuter experience. Made available in the Sanntid app, this information allows users to make informed decisions about their route choice, travel time, and travel mode alternatives (Kolumbus, 2023). Real-time travel information systems are common solutions, but what makes the Sanntid app stand out from similar options like Google Maps is its integrations with the city bikes system, offering live updates on bike availability, bike charge levels, and nearest city bike stations. Aggregating this information in one app simplifies the use of the application and facilitates multimodal transport.

The HJH initiative is a rather successful initiative for promoting the use of the bike-sharing system (Pritchard & Lovelace, 2022). However, it remains unclear whether users choose cycling as their primary mode of transport or still rely on public transport for their daily commute and use the bikes occasionally (Semenescu & Coca, 2022). The HJH initiative has also untapped potential to integrate feedback and self-monitoring and social networks, under the assumption that these functions could help people in the same workplaces to motivate each other's commute cycling. Our assessment revealed that the health project initiative received a positive response from the participants and thus may help promote cycling to work (HjemJobbHjem, 2024b). However, the limited number of participants and short campaign time of the project make extending its results to a greater population difficult.

Among Stavanger's smart mobility initiatives, the Sanntid app and the HJH scheme provide considerable developments in encouraging multimodal transport and cycling. However, possible long-term effects on travel behaviours of these initiatives should be assessed thoroughly, and further research and data-driven evaluations are necessary to evaluate whether various initiatives can reduce the city's dependency on private car use.

Addressing issues related to cycling infrastructure (Buck & Buehler, 2012; Fishman et al., 2012; Noland et al., 2016), safety (Fishman, 2016; Rérat, 2019; Ricci, 2015; Teixeira et al., 2021), social equity, and sustainability (Poslad et al., 2015; Sunio & Schmöcker, 2017; L. Wang, 2018) will be crucial for realizing the full potential of initiatives to promote (commute) cycling beyond smart mobility solutions. Despite multiple uncertainties, regarding overall effectiveness in boosting cycling rates, local policymakers need to assess, develop, support, and protect any promising initiatives and related smart solutions in Stavanger to make cycling more attractive to its citizens.

Conclusion

Like other aspects of a smart city, smart mobility comes with advantages and disadvantages. While many of the smart mobility solutions discussed in this chapter aim to make commuting by bike more affordable and accessible, some issues such as data privacy, particularly the collection of users' live location data, remain a significant challenge that needs to be addressed.

The results of this study provide insights for future policymaking, not just in Stavanger but in other cities looking to promote sustainable transportation alternatives. This study has depicted relationships between the initiatives to promote commute (cycling) and various types of smart solutions for the cycling sector. This knowledge is critical for policymakers to anticipate policy strategies focusing on certain smart solutions and various target groups.

References

Aderibigbe, O.-O., & Gumbo, T. (2024). Smart cities and their impact on urban transportation systems and development. In *Emerging technologies for smart cities: Sustainable transport planning in the Global North and Global South* (pp. 105–129). Springer.

Albuquerque, V., Sales Dias, M., & Bacao, F. (2021). Machine learning approaches to bike-sharing systems: A systematic literature review. *ISPRS International Journal of Geo-Information*, 10(2), 62.

Ayad, L., Imine, H., Lantieri, C., & De Crescenzio, F. (2024). Pedal towards safety: The development and evaluation of a risk index for cyclists. *Infrastructures*, 9(1), 14.

Bachand-Marleau, J., Lee, B., & El-Geneidy, A. (2012). Better understanding of factors influencing likelihood of using shared bicycle systems and frequency of use. *Transportation Research Record*, 2314, 66–71. doi:10.3141/2314-09

Balakrishna, R., Ben-Akiva, M., Bottom, J., & Gao, S. (2013). Information impacts on traveler behavior and network performance: State of knowledge and future directions. In S. Ukkusuri & K. Ozbay (Eds.), *Advances in dynamic network modeling in complex transportation systems* (Vol. 2, pp. 193–224). Springer.

Ball, K., Hunter, R.F., Maple, J.-L., Moodie, M., Salmon, J., Ong, K.-L., Stephens, L.D., Jackson, M., & Crawford, D. (2017). Can an incentive-based intervention increase physical activity and reduce sitting among adults? The ACHIEVE (Active Choices IncEntiVE) feasibility study. *International Journal of Behavioral Nutrition and Physical Activity*, 14, 1–10.

Batterbury, S. (2003). Environmental activism and social networks: Campaigning for bicycles and alternative transport in West London. *The Annals of the American Academy of Political and Social Science*, 590(1), 150–169.

Bearne, S. (2021). *Is it worth tracking your carbon footprint?* BBC. https://www.bbc.com/news/business-55907643

Ben-Elia, E., & Ettema, D. (2009). Carrots versus sticks: Rewarding commuters for avoiding the rush-hour – A study of willingness to participate. *Transport Policy*, 16(2), 68–76.

Benevolo, C., Dameri, R.P., & D'auria, B. (2016). Smart mobility in smart city: Action taxonomy, ICT intensity and public benefits. In T. Torre, A. Braccini, & R. Spinelli (Eds.), *Empowering organizations: Enabling platforms and artefacts* (pp. 13–28). Springer International.

Binsted, A., & Hutchins, R. (2012). *The role of social networking sites in changing travel behaviours.* TRL.

Bliemer, M.C., & van Amelsfort, D.H. (2008). Rewarding instead of charging road users: A model case study investigating effects on traffic conditions. *European Tranport/Trasporti Europei*, 44, 23–40.

Böcker, L., Dijst, M., Faber, J., & Helbich, M. (2015). En-route weather and place valuations for different transport mode users. *Journal of Transport Geography*, 47, 128–138.

Bonneau, J., & Preibusch, S. (2010). The privacy jungle: On the market for data protection in social networks. In *Economics of information security and privacy* (pp. 121–167). Springer.

Brčić, D., Slavulj, M., Šojat, D., & Jurak, J. (2018, May 17–19). *The role of smart mobility in smart cities* [Paper presentation]. Fifth International Conference on Road and Rail Infrastructure (CETRA 2018). Zadar, Croatia.

Brinkhoff, T. (2024). *City population, Stavanger*. https://www.citypopulation.de/en/norway/admin/rogaland/1103__stavanger/

Buchanan, K., Russo, R., & Anderson, B. (2014). Feeding back about eco-feedback: How do consumers use and respond to energy monitors? *Energy Policy*, 73, 138–146.

Buck, D., & Buehler, R. (2012, January 22–26). *Bike lanes and other determinants of capital bikeshare trips* [Paper presentation]. 91st Transportation Research Board annual meeting, Washington, DC, USA.

Cellina, F., Castri, R., Simão, J.V., & Granato, P. (2020). Co-creating app-based policy measures for mobility behavior change: A trigger for novel governance practices at the urban level. *Sustainable Cities and Society*, 53, 101911.

Cerutti, P.S., Martins, R.D., Macke, J., & Sarate, J.A.R. (2019). 'Green, but not as green as that': An analysis of a Brazilian bike-sharing system. *Journal of Cleaner Production*, 217, 185–193.

Cervero, R., & Duncan, M. (2003). Walking, bicycling, and urban landscapes: Evidence from the San Francisco Bay Area. *American Journal of Public Health*, 93(9), 1478–1483.

Chen, S.-Y., & Lu, C.-C. (2016). A model of green acceptance and intentions to use bike-sharing: YouBike users in Taiwan. *Networks and Spatial Economics*, 16, 1103–1124.

Chen, Z., van Lierop, D., & Ettema, D. (2020). Exploring dockless bikeshare usage: A case study of Beijing, China. *Sustainability*, 12(3), 1238.

Cho, E., Myers, S.A., & Leskovec, J. (2011). Friendship and mobility: User movement in location-based social networks. In V.A. Chidanand, J. Ghosh, & P. Smyth (Eds.), *KDD '11: The 17th ACM SIGKDD International Conference on Knowledge Discovery and Data Mining San Diego California USA August 21 - 24, 2011* (pp. 1082–1090). Association for Computing Machinery. doi:10.1145/2020408.202057

Dziekan, K., & Kottenhoff, K. (2007). Dynamic at-stop real-time information displays for public transport: Effects on customers. *Transportation Research Part A: Policy and Practice*, 41(6), 489–501.

Eckhoff, D., & Wagner, I. (2017). Privacy in the smart city – Applications, technologies, challenges, and solutions. *IEEE Communications Surveys & Tutorials*, 20(1), 489–516.

Eiterjord, G., Garpe, L., & Berre, H. (2011, December 27). Flere syklister – Og vi er utålmodige! *Stavanger Aftenblad*. https://www.aftenbladet.no/meninger/i/PaQRb/flere-syklister-og-vi-er-utaalmodige

Festinger, L. (1954). A theory of social comparison processes. *Human Relations*, 7(2), 117–140.

Fishman, E. (2016). *Cycling as transport* (Vol. 36). Taylor & Francis.

Fishman, E., Washington, S., & Haworth, N. (2012). Barriers and facilitators to public bicycle scheme use: A qualitative approach. *Transportation Research Part F: Traffic Psychology and Behaviour*, 15(6), 686–698.

Fishman, E., Washington, S., & Haworth, N. (2014). Bike share's impact on car use: Evidence from the United States, Great Britain, and Australia. *Transportation Research Part D: Transport and Environment*, 31, 13–20.

Fishman, E., Washington, S., Haworth, N., & Mazzei, A. (2014). Barriers to bikesharing: An analysis from Melbourne and Brisbane. *Journal of Transport Geography*, 41, 325–337.

Fishman, E., Washington, S., Haworth, N., & Watson, A. (2015). Factors influencing bike share membership: An analysis of Melbourne and Brisbane. *Transportation Research Part A: Policy and Practice*, 71, 17–30.

Giffinger, R., Fertner, C., Kramar, H., Kalasek, R., Pichler-Milanovic, N., & Meijers, E.J. (2007). *Smart cities. Ranking of European medium-sized cities.* Final report. Institut für Raumplanung.

Gonçalves, G.d.L., Neiva, S.d.S., Deggau, A.B., de Oliveira Veras, M., Ceci, F., de Lima, M.A., & de Andrade Guerra, J.B.S.O. (2021). The impacts of the fourth industrial revolution on smart and sustainable cities. *Sustainability*, 13(13), 1–21.

Gössling, S. (2013). Urban transport transitions: Copenhagen, city of cyclists. *Journal of Transport Geography*, 33, 196–206.

Government.no. (2021). *National transport plan 2022–2033.* https://www.regjeringen. no/en/topics/transport-and-communications/content-2021/national-transport-pla n-20222033/id2866098/

Heinen, E., & Handy, S. (2012). Similarities in attitudes and norms and the effect on bicycle commuting: Evidence from the bicycle cities Davis and Delft. *International Journal of Sustainable Transportation*, 6(5), 257–281.

Heinen, E., Van Wee, B., & Maat, K. (2010). Commuting by bicycle: An overview of the literature. *Transport Reviews*, 30(1), 59–96.

HjemJobbHjem. (2019). *Helseprosjekt i mål.* https://www.kolumbus.no/aktuelt/helsep rosjekt-i-mal/.

HjemJobbHjem. (2021). *Mobility solutions for employees in the Nord-Jæren region.* Stavanger Municipality.

HjemJobbHjem. (2024a). *Bedriftsavtalen.* https://www.hjemjobbhjem.no/rogaland/bed riftsavtalen/

HjemJobbHjem. (2024b). *Jobbreisen til bedre helse.* https://www.hjemjobbhjem.no/rogala nd/gode-historier/jobbreisen-til-bedre-helse/

Hosford, K., Lear, S.A., Fuller, D., Teschke, K., Therrien, S., & Winters, M. (2018). Who is in the near market for bicycle sharing? Identifying current, potential, and unlikely users of a public bicycle share program in Vancouver, Canada. *BMC Public Health*, 18, 1–10.

Hosford, K., & Winters, M. (2018). Who are public bicycle share programs serving? An evaluation of the equity of spatial access to bicycle share service areas in Canadian cities. *Transportation Research Record*, 2672(36), 42–50.

Huovila, A., Bosch, P., & Airaksinen, M. (2019). Comparative analysis of standardized indicators for smart sustainable cities: What indicators and standards to use and when? *Cities*, 89, 141–153.

Ismagilova, E., Hughes, L., Rana, N.P., & Dwivedi, Y.K. (2022). Security, privacy and risks within smart cities: Literature review and development of a smart city interaction framework. *Information Systems Frontiers*, 24, 393–414.

Iwińska, K., Blicharska, M., Pierotti, L., Tainio, M., & de Nazelle, A. (2018). Cycling in Warsaw, Poland – Perceived enablers and barriers according to cyclists and non-cyclists. *Transportation Research Part A: Policy and Practice*, 113, 291–301.

Jakicic, J.M., Davis, K.K., Rogers, R.J., King, W.C., Marcus, M.D., Helsel, D., Rickman, A.D., Wahed, A.S., & Belle, S.H. (2016). Effect of wearable technology combined with a lifestyle intervention on long-term weight loss: The IDEA randomized clinical trial. *JAMA*, 316(11), 1161–1171.

Kazhamiakin, R., Marconi, A., Perillo, M., Pistore, M., Valetto, G., Piras, L., Avesani, F., & Perri, N. (2015, October 25–28). *Using gamification to incentivize sustainable urban mobility* [Paper presentation]. 2015 IEEE First International Smart Cities Conference (ISC2). Guadalajara, Mexico.

Kolumbus. (2019). *Real-time app: Kolumbus Sanntid.* https://www.kolumbus.no/en/useful-information/apps/real-time-kolumbus-sanntid/

Kolumbus. (2021a). *The city bike.* https://www.kolumbus.no/en/travel/bicycle/bysykkelen/

Kolumbus. (2021b). *Vilkår for bruk av Stiim-appen.* https://www.kolumbus.no/hjelp-og-kontakt/reisevilkar/vilkar-for-bruk-av-stiim-appen/

Kolumbus. (2022a). *Kolumbus for a smarter city.* https://www.kolumbus.no/en/news-archive/kolumbus-for-a-smarter-city/

Kolumbus. (2022b). *Smart mobility and public transportation initiatives in Stavanger.*

Kolumbus. (2023). *Sanntidssystemet.* https://www.kolumbus.no/reise/kart/sanntidskart/sanntidssystemet/

Kolumbus. (2024a). *Priser og billetter.* https://www.kolumbus.no/Billetter/-priser-og-produkter/

Kolumbus. (2024b). *Stiim-app-and-privacy-policy.* https://www.kolumbus.no/en/help-and-contact/rules-and-conditions/privacy-policy/stiim-app-and-privacy-policy/

Kramers, A., Höjer, M., Lövehagen, N., & Wangel, J. (2014). Smart sustainable cities – Exploring ICT solutions for reduced energy use in cities. *Environmental Modelling & Software*, 56, 52–62.

Krizek, K.J., Barnes, G., & Thompson, K. (2009). Analyzing the effect of bicycle facilities on commute mode share over time. *Journal of Urban Planning and Development*, 135 (2), 66–73. doi:10.1061/(ASCE)0733–9488(2009)135:2(66)

Kurauchi, F., Wahaballa, A.M., Othman, A.M., Uno, N., & Takagi, A. (2019). Determinants of travel choice behaviour with travel-time variability in the presence of real-time information. *International Journal of Intelligent Transportation Systems Research*, 17, 61–73.

Lättman, K., & Otsuka, N. (2024). *Sustainable development of urban mobility through active travel and public transport* (Vol. 16). MDPI.

Lee, K., & Sener, I.N. (2021). Strava Metro data for bicycle monitoring: A literature review. *Transport Reviews*, 41(1), 27–47.

Mackett, R.L. (2014). The health implications of inequalities in travel. *Journal of Transport & Health*, 1(3), 202–209.

Maldonado Silveira Alonso Munhoz, P.A., da Costa Dias, F., Kowal Chinelli, C., Azevedo Guedes, A.L., Neves dos Santos, J.A., da Silveira e Silva, W., & Pereira Soares, C.A. (2020). Smart mobility: The main drivers for increasing the intelligence of urban mobility. *Sustainability*, 12(24), 10675.

Manca, F., Daina, N., Sivakumar, A., Yi, J.W.X., Zavitsas, K., Gemini, G., Vegetti, I., Dargan, L., & Marchet, F. (2022). Using digital social market applications to incentivise active travel: Empirical analysis of a smart city initiative. *Sustainable Cities and Society*, 77, 103595.

Mora, L., Bolici, R., & Deakin, M. (2017). The first two decades of smart-city research: A bibliometric analysis. *Journal of Urban Technology*, 24(1), 3–27.

Müller-Eie, D., Knutsen, E., & Selland, E.P. (2019). Personal travel planning: Review of soft transport measure criteria and effects. *WIT Transactions on Ecology and the Environment*, 238, 603–613.

Murphy, E., & Usher, J. (2015). The role of bicycle-sharing in the city: Analysis of the Irish experience. *International Journal of Sustainable Transportation*, 9(2), 116–125.

Nasjonal reisevaneundersøkelse. (2021). *Nasjonal reisevaneundersøkelse (RVU) nøkkeltallsrapport 2020.* https://www.vegvesen.no/globalassets/fag/fokusomrader/na

sjonal-transportplan-ntp/reisevaner/2020/nokkeltallsrapport-2020-versjon-per-20.12.
21.pdf

Netpower. (2019). *Unique expertise in mobility solutions.* https://www.netpower.no/uni
k-ekspertise-pa-mobilitetslosninger/

Noland, M., Moran, T., & Kotschwar, B.R. (2016). *Is gender diversity profitable? Evidence
from a global survey* (Peterson Institute for International Economics Working Paper 16-13).

Paiva, S., Ahad, M.A., Tripathi, G., Feroz, N., & Casalino, G. (2021). Enabling technologies
for urban smart mobility: Recent trends, opportunities and challenges. *Sensors*, 21(6), 2143.

Pan, L., Xia, Y., Xing, L., Song, Z., & Xu, Y. (2022). Exploring use acceptance of electric bicycle-
sharing systems: An empirical study based on PLS-SEM analysis. *Sensors*, 22(18), 7057.

Pan, M.L. (2016). *Preparing literature reviews: Qualitative and quantitative approaches.*
Routledge.

Pasca, M.G., Guglielmetti Mugion, R., Toni, M., Di Pietro, L., & Renzi, M.F. (2021).
Gamification and service quality in bike sharing: An empirical study in Italy. *The TQM
Journal*, 33(6), 1222–1244.

Poslad, S., Ma, A., Wang, Z., & Mei, H. (2015). Using a smart city IOT to incentivise and
target shifts in mobility behaviour – Is it a piece of pie? *Sensors*, 15(6), 13069–13096.
doi:10.3390/s150613069.

Pritchard, R., & Lovelace, R. (2022). *Sykkelpotensial og bysykler En beregning av potensialet
for økt hverdagssykling og evaluering av bysykkelordningene på Nord-Jæren, i Trondheim og
i Bergen.* NORCE Helse og samfunn.

Pucher, J., Dill, J., & Handy, S. (2010). Infrastructure, programs, and policies to increase
bicycling: An international review. *Preventive Medicine*, 50, S106–S125.

Raux, C., Zoubir, A., & Geyik, M. (2017). Who are bike sharing schemes members and do
they travel differently? The case of Lyon's 'Velo'v' scheme. *Transportation Research Part
A: Policy and Practice*, 106, 350–363.

Reilly, K.H., Noyes, P., & Crossa, A. (2020). From non-cyclists to frequent cyclists: Factors associated with frequent bike share use in New York City. *Journal of Transport &
Health*, 16, 100790.

Rérat, P. (2019). Cycling to work: Meanings and experiences of a sustainable practice.
Transportation Research Part A: Policy and Practice, 123, 91–104.

Rérat, P. (2021). The rise of the e-bike: Towards an extension of the practice of cycling?
Mobilities, 16(3), 423–439. doi:10.1080/17450101.2021.1897236

Ricci, M. (2015). Bike sharing: A review of evidence on impacts and processes of implementation and operation. *Research in Transportation Business & Management*, 15, 28–38.

Rozas, L.W., & Klein, W.C. (2010). The value and purpose of the traditional qualitative
literature review. *Journal of Evidence-Based Social Work*, 7(5), 387–399.

Semenescu, A., & Coca, D. (2022). Why people fail to bike the talk: Car dependence as a
barrier to cycling. *Transportation Research Part F: Traffic Psychology and Behaviour*, 88,
208–222.

Skoglund, J. (2019). *Stavanger is the country's smartest in transport – Received the Mobility
Award 2019.* https://www.veier24.no/artikler/stavanger-karet-til-landets-smarteste-innen-
transport-fikk-mobilitetsprisen-2019/471361

Stavanger Municipality. (2018). *Stavanger climate and environmental plan 2018–2030.*
https://www.stavanger.kommune.no/siteassets/renovasjon-klima-og-miljo/miljo-og-k
lima/climate-and-environmental-plan-stavanger-2018-2030—final-version.pdf

Stavanger Municipality. (2024a). *Hovedrutene for sykkel.* https://www.stavanger.komm
une.no/vei-og-trafikk/stavanger-pa-sykkel/hovedrutene-for-sykkel/

Stavanger Municipality. (2024b). *Sykkelprioriterte gater.* https://www.stavanger.komm une.no/vei-og-trafikk/stavanger-pa-sykkel/sykkelprioriterte-gater/

Sunio, V., & Schmöcker, J.-D. (2017). Can we promote sustainable travel behavior through mobile apps? Evaluation and review of evidence. *International Journal of Sustainable Transportation,* 11(8), 553–566.

Šurdonja, S., Giuffrè, T., & Deluka-Tibljaš, A. (2020). Smart mobility solutions – Necessary precondition for a well-functioning smart city. *Transportation Research Procedia,* 45, 604–611.

Swan, M. (2012). Sensor mania! The Internet of Things, wearable computing, objective metrics, and the quantified self 2.0. *Journal of Sensor and Actuator Networks,* 1(3), 217–253.

Taranger, I.S., Bogen, E., & Otterdal, Ø. (2023). *Ingen vil betale ekstraregning for sykkelvei til 2,2 mrd.* NRK. https://www.nrk.no/rogaland/uenighet-om-hvem-som-skal-beta le-for-sykkelstamvei-som-blir-dyrere-og-dyrere-1.16617838

Teixeira, J.F., Silva, C., & Moura e Sá, F. (2021). The motivations for using bike sharing during the COVID-19 pandemic: Insights from Lisbon. *Transportation Research Part F: Traffic Psychology and Behaviour,* 82, 378–399. doi:10.1016/j.trf.2021.09.016

Thogersen, J. (2007). Social marketing of alternative transportation modes. In *Threats from car traffic to the quality of urban life: Problems, causes and solutions* (pp. 367–381). Emerald.

Wang, L. (2018). Barriers to implementing pro-cycling policies: A case study of Hamburg. *Sustainability,* 10(11), 4196.

Wang, W., Gan, H., Wang, X., Lu, H., & Huang, Y. (2022). Initiatives and challenges in using gamification in transportation: A systematic mapping. *European Transport Research Review,* 14(1), 41.

Wei, T., Wu, J., & Chen, S. (2021). Keeping track of greenhouse gas emission reduction progress and targets in 167 cities worldwide. *Frontiers in Sustainable Cities,* 3, 696381.

White, T., & Marchet, F. (2021). Digital social markets: Exploring the opportunities and impacts of gamification and reward mechanisms in citizen engagement and smart city services. In K.P. Valavanis (Ed.), *How smart is your city? Technological innovation, ethics and inclusiveness* (pp. 103–125). Springer.

Wu, C., Kim, I., & Chung, H. (2021). The effects of built environment spatial variation on bike-sharing usage: A case study of Suzhou, China. *Cities,* 110, 103063.

Yang, L., Sahlqvist, S., McMinn, A., Griffin, S.J., & Ogilvie, D. (2010). Interventions to promote cycling: Systematic review. *BMJ,* 341, 1–10.

Zhao, J., Wang, J., Xing, Z., Luan, X., & Jiang, Y. (2018). Weather and cycling: Mining big data to have an in-depth understanding of the association of weather variability with cycling on an off-road trail and an on-road bike lane. *Transportation Research Part A: Policy and Practice,* 111, 119–135.

14 Developing children's understanding of their complex urban environment – kindergarten's awareness of air quality in Stavanger

Barbara Maria Sageidet, Todor Milkov Kesarovski and Petar Zhivkov

Introduction

Urbanization and densification of human activities have been recognized as processes that tend to pose challenges to air quality due to the spatial concentrations of heavier vehicular traffic and industries, both globally (Baklanov et al., 2016; Borck & Schrauth, 2021; Liang & Gong, 2020) and within Norway (Castell et al., 2018; Norwegian Institute of Public Health [NIPH], 2018). The World Health Organization (WHO, 2021) emphasizes the close relationship between air pollutants and increased morbidity or mortality. Consequently, there is a growing recognition of the worrisome association between air quality and health (Castell et al., 2018; Kopnina, 2017; WHO, 2006), and particularly children's heightened vulnerability (Bennett et al., 1996; Bettiol et al., 2021; Garcia et al., 2021). International research reports of air pollution challenges for citizens include children attending kindergartens as a vulnerable group (Liu et al., 2013; Rehwagen et al., 1999), with Norway also beginning to acknowledge these issues (Castell et al., 2018). In Norway, outdoor education is an integral part of kindergarten activities (Sageidet et al., 2018), further highlighting the importance of understanding and addressing local air quality concerns.

Worldwide, air pollution is a major cause of health problems and diseases (WHO, 2021). Amongst the most vulnerable groups are children, whose respiratory systems are not fully developed and whose airway calibre is narrower. Consequently, children experience greater particle deposition of irritants in the tracheobronchial region compared to adults, potentially leading to more significant effects (Bennett et al., 1996; Bettiol et al., 2021; Garcia et al., 2021). Furthermore, due to their higher body surface area–to-weight ratio, children have increased potential for exposure and absorption of pollutants (Bennett et al., 1996; Garcia et al., 2021). For these and other reasons, like susceptibility to asthma or other diseases, children may suffer adverse effects from lower levels of pollutants than those affecting adults. Despite the greater risk of acute respiratory illness in children in larger cities (Rodriguez et al., 2007), the negative impact of air pollution on children's health remains a significant concern everywhere, also in Stavanger (Stavanger Municipality, 2018).

DOI: 10.4324/9781003498650-18

Air pollution is also linked to climate change challenges, as carbon dioxide and air pollutants are co-emitted from fossil fuel combustion (Zhong et al., 2023). Special attention is being paid to air quality to reduce the adverse per capita environmental impact of cities to achieve global Sustainable Development Goal (SDG) 11 on sustainable cities and human settlements (United Nations, 2015, Target 11.6). Stavanger is one of 112 European cities with the ambition to become climate neutral by 2030 (European Commission, 2024; Gaglione, 2023).

Within this thematic framework, this chapter explores kindergartens' air quality awareness in Stavanger, Norway. Specifically, the chapter will explore the research questions: how aware are early childhood teachers and their kindergartens of the air quality in their nearby outside areas? How does air quality influence the kindergartens' and the early childhood teachers' plan for outdoor activities with the children? How do early childhood teachers communicate air quality and its environmental role with the children?

For this purpose, the chapter reviews how air quality monitoring is performed in Stavanger. Further, it qualitatively explores the literature, early childhood teachers' answers to a qualitative questionnaire, and two interviews with professionals from the municipality. By drawing on theories related to urban air quality, children's health, and environmental citizenship, the chapter uncovers untapped potentials in both citizens' and local kindergartens' awareness and understanding of air quality issues. It concludes by offering suggestions for fostering children's early comprehension of the interconnected aspects of their urban environment.

Research background and theory

Air pollutants and their effects on public health

Air pollution consists of a range of different substances, depending on the source, with particulate matter (PM), nitrogen dioxide (NO_2), and ozone (O_3) being the most significant contributors in urban areas today (NIPH, 2018). Among these, PM stands out as a major concern due to its adverse health effects (Bettiol et al., 2021; Garcia et al., 2021; Yu & Zhang, 2023). PM exists in the form of tiny particles and liquid droplets dispersed in the atmosphere, varying in size, shape, surface area, chemical composition, and origin. Sources of PM include transportation; energy production; industrial, commercial, and residential activities; biomass burning; and agriculture, with traffic and combustion being primary contributors in urban areas (IQAir, 2020; Squizzato et al. 2017). Notably, scientific research does not indicate a safe level of exposure or a threshold below which no adverse health effects occur (WHO, 2021). In Europe, where more than 80% of the population live in cities exceeding WHO's air quality guidelines, PM pollution is associated with a reduction in life expectancy by almost 9 months on average (WHO, 2006). Furthermore, Yu and Zhang (2023) discovered that higher concentrations of daily $PM_{2.5}$ and PM_{10} air pollution are linked to decreased physical activity and increased sedentary behaviour in children. Despite a slight decrease in average concentrations over the last decade, PM levels in cities remain worrisome (WHO, 2021).

Nitrogen dioxide (NO_2) belongs to nitrogen oxides (NO_x). It is an indicator of this large group of reactive gases (Greenwood & Earnshaw, 1997). NO_2 accumulates in the air from combustion engines burning fossil fuels in motor vehicles, power plants, and off-road equipment (NIPH, 2018). Short-term exposure to NO_2 may aggravate respiratory diseases like asthma and lead to coughing, wheezing, difficulty breathing, or even medical needs. Longer exposures to higher concentrations of NO_2 may contribute to the development of asthma and increase susceptibility to respiratory infections (Holst et al., 2020). Children, people with asthma, and the elderly are particularly at risk of NO_2-related health effects (WHO, 2006, 2021).

Monitoring and public dissemination of data on air quality in Stavanger

As awareness of the risks posed by urban air pollution grows, efforts to monitor and share air quality data have advanced significantly. In Norway, ambient air quality data on PM_{10}, $PM_{2.5}$, and NO_2 are collected from stationary monitoring stations and collected in a central public database, accessible under the Norwegian Licence for Open Government Data. This enables real-time data use by various entities for analysis, aggregation, or dissemination. A large number of stations in the air quality monitoring network are positioned along busy roads (i.e., traffic stations) to cover areas where the highest concentrations occur, while others serve as urban background stations, located away from direct pollution sources. Stavanger hosts four monitoring stations: two traffic stations at Kannik and Schancheholen and two urban background stations at Våland and Vågen (see Figure 14.1). Although annual mean values suggest relatively stable air pollution levels for 2010 to 2020 (see Table 14.1), occasional spikes can be observed on a weekly, daily, or hourly basis (https://luftkvalitet.nilu.no/;Schneider et al., 2023).

Figure 14.1 Air quality monitoring stations in the municipality of Stavanger
Source: Basemap Copyright Credit: 2020 Esri and its licensors. All rights reserved.

Table 14.1 Annual mean levels of critical air pollutants in Stavanger, Norway

Pollutant	2010	2011	2012	2013	2014	2015	2016	2017	2018	2019	2020
NO_2 ($\mu g/m^3$)	—	52	45	44	41	37	33	32	28	25	22
PM_{10} ($\mu g/m^3$)	29	27	26	28	24	28	22	13	14	13	16
$PM_{2.5}$ ($\mu g/m^3$)	12	10	8	10	10	10	10	8	9	9	8

Source: Central database for local air quality (SDB) via NILU and Miljøstatus.

Robust monitoring systems to track pollution levels and trends are fundamental to effective air quality management. However, disseminating these data is equally vital to safeguarding public health and vulnerable groups. Research underscores the necessity of communicating air quality information, especially during pollution peaks, because many individuals lack a comprehensive understanding of the health impacts of air pollution (Maione et al., 2021; Wu et al., 2021). Thus, efforts must be made to present air quality data in an understandable, jargon-free manner to ensure comprehension among non-experts. The public can be effectively informed about air quality and associated health hazards in real time using data streams, contemporary information, and communication technologies (Davila et al., 2015).

Some cities offer special guidance to pollution-sensitive populations, like children, the elderly, pregnant women, and those with pre-existing health conditions, on modifying behaviour and mobility during high-pollution days through traffic signs, web-based announcements, and radio broadcasts (Stauffer, 2017). Other cities include regional warning or alert systems that send personalized messages to registered participants; for example, specialized notification apps in Amsterdam and Utrecht (Netherlands), smog alerts with tailored action advice in Antwerp (Belgium), and voice alarms, text messaging, email, or website communication in Skawina, Pruszków, and Katowice (Poland). Additionally, Helsinki's pharmacists distribute informational materials during peak street dust levels, while Rennes (France) and Iasi (Romania) raise pollution awareness through messaging schools and daycare facilities. In Norway, regardless of where they live, people can subscribe to an alert system for information on air pollution levels, and air quality levels are also integrated into the yr.no app[1] alongside weather updates. Overall, these initiatives aim to deliver timely and relevant air quality information to the public through smart city approaches, enabling concerned citizens to make informed decisions to protect their health (Stauffer, 2017).

Environmental citizenship

Environmental citizenship builds on social constructionist theories (Vygotzky, 1986) that underline the importance of practical activities and social contexts to promote learning amongst children and adults, including the development of skills and knowledge essential for sustainability transitions (Bell, 2016).

Environmental citizenship is related to global citizenship, a political, ecological, economic, and social way of understanding and acting, based on universal values, responsibility, active participation, and critical thinking, and underlines humans as members of the ecological system of the planet, together with all other organisms (Lee & Fouts, 2005; Sageidet & Heggen, 2021; United Nations, 2015). Children's environmental development is associated with outdoor experiences and related learning activities (Heggen et al. 2019). Thus, children with an active identity as ecological (or environmental) citizens may feel a sense of belonging to their local place and to our planet and may develop a desire of care for it, as well as solidarity, curiosity, and knowledge (Sageidet & Heggen, 2021). DOGA et al. (2019) envision health-promoting cities ensuring citizens the right to clean air, clean water, and green recreation areas. According to the Norwegian law from May 13, 2014,

> Everyone has the right to an environment that ensures health. *[…]* Citizens have the right to get knowledge about the state of the natural environment and about the effects of planned and implemented interventions in nature, so that they can safeguard the right they have in line with the previous subsection.
> (The Constitution of the Kingdom of Norway, 2014, § 112)

Children's learning about and for the environment

Internationally, there is an increasing focus on children's learning about and agency for the environment (Heggen et al., 2019; Kopnina, 2017). Kos et al. (2016) interviewed 5- to 6-year-old children and found an initially low understanding of relationships, such as walking instead of driving and the respective effect on the environment. This limited comprehension may be explained by the prevalent use of cars – approximately 66% for escorting children to kindergartens in Stavanger, as observed by Hernández-Palacio and Kesarovski (2024). However, Kos et al. found that the children developed their understandings during their involvement in activities in which they gained related background information. Conversations with 4- to 5-year-old children in kindergartens in Queensland, Australia, and Rogaland, Norway (Sageidet et al., 2019), as well as drawings by 3- to 6-year-old children in Turkey (Duran, 2021), revealed limited but significant understanding of sustainability-related topics and some sustainability-related relationships.

Stavanger was among the first Norwegian municipalities to encourage kindergartens to work systematically on projects focused upon sustainable development and environmental protection. A public kindergarten in Stavanger was the second one in Norway to achieve the international environmental 'green flag' certification, in 2002. This label is associated with the international Foundation for Environmental Education, FEE (Jupskås & Römcke, 2002; see also Stavanger Municipality, 2018). In 2019–2020, children were included in the planning of the smart city project Lervig smart park, located in a dense residential area east of the Stavanger city centre, in line with a Norwegian smart cities' focus on citizen participation (DOGA et al., 2019; Montalvan Castilla & Müller, 2023).

Methodology

Using interdisciplinary inquiry (Derry et al., 2014), this study examines air quality monitoring in Stavanger on the one hand and explores the awareness and implications of it among citizens, specifically kindergartens, on the other hand. The study aims to promote dialogue, understanding, and knowledge.

Within a descriptive qualitative research design (Mayring, 2014), the study includes two semi-structured interviews with professionals in the municipality and a small survey of subjective evaluations of early childhood teachers including the age group of children between 3 and 6 years in Stavanger. The semi-structured questionnaire with open- and closed-ended questions (Beckett & Clegg, 2007) included information about the study and the optional consent and was approved by the Norwegian Agency for Shared Services in Education and Research.

The informants were asked about their awareness, interests, and/or concerns related to air quality in the vicinity of their kindergartens in Stavanger. They were further asked about their knowledge of air quality in Stavanger, including NO_2 and particulate matter ($PM_{2.5}$ and PM_{10}), and how they may deal with air quality and pollution in their roles and work within the kindergarten. They were asked whether air quality influences the planning and implementation of activities with the children outdoors in the vicinity of the kindergarten and whether air quality is a theme they focus on together with the children and, if so, how. Finally, the early childhood teachers were asked about parents' usual transport choices between home and kindergarten and how they evaluate parents' interests and/or concerns related to air quality.

The answers to the questionnaires and the content of the interviews were analyzed using qualitative content analysis (Mayring, 2014). The study was undertaken with an awareness of the researchers' preunderstandings and their professional perspectives and personal biases, including (unconscious) expectations and beliefs, which may influence the reflective act of interpreting (Mayring, 2014). The findings were discussed considering the presented theory.

Data collection process

From April 2022 to February 2024, 32 kindergartens listed on Stavanger Municipality's webpage were asked to participate in the survey. Most were contacted by telephone before or after sending an e-mail request. Reminders were sent by e-mail up to three times to increase the chance of participation.

Seven informants returned a completed questionnaire, which is a low response rate. Many kindergartens refused to participate, due to lack of time, an intense work situation, illness among the staff, or involvement in other research. All data from the returned questionnaires were handled confidentially and individual identifiers were removed ahead of the analysis. The interview informants kindly agreed to be identified in this chapter.

The responders

Among the seven responding early childhood teachers, five were women and their age range was between 37 and 51 years. All but one had experience as a leader or were leaders of their kindergarten. Their years of work experience was between 18 and 27 years, with the exception of one informant with 5 years of experiences. The informants had worked in their current kindergarten for 2 to 18 years.

Two interviews were conducted at the Department of Environmentally Oriented Health Care in Stavanger and Rogaland County, which is nationally obligated to survey the air quality and provide supervision. A short interview was held with Cathrine Telstø Pedersen, the leader of that department, on behalf of the health supervision in the county. In this position for only 6 weeks, she has 18 years of earlier work experience as an early childhood teacher. The main interview was held with Geir Tore Aamdal, the scientific advisor for environmentally oriented health care at the department. He has an educational background within nature, health, and environment and has 22 years of professional experience within environmentally oriented health care and 10 years of experience particularly working with air quality. Aamdal is responsible for the implementation of the environmental protection plan for Stavanger, monitoring and surveying the air quality, and quality assurance of monitoring data, reporting, and administration.

Results and discussion

The low response rate to the questionnaire might indicate that air quality is not a high-priority interest among early childhood teachers, and maybe this is similar among other groups of citizens. In fact, the air quality in Stavanger seems to stay relatively stable (Schneider et al., 2023), and most children in this medium-sized city have access to urban park areas and gardens –even natural areas are not too far from the city centre (Sageidet et al., 2018). However, the slight correlation between urban population densities and air pollution in Stavanger (Zhivkov & Kesarovski, 2024), together with stronger correlations in other European cities (Kopnina, 2017; Rehwagen et al., 1999; Zhong et al., 2023), implies that kindergartens near the city centre, like the ones in this study, have a higher risk of reduced outdoor air quality than rural kindergartens.

Three of the early childhood teachers in the study indicated a medium level of interest/concern about air quality in the vicinity of their kindergarten, while three others expressed low interest/concern. Their responses regarding the kindergarten's interest as an institution on the topic were nearly identical. However, the informants, most of whom (five of seven) are also leaders of their kindergarten, seem to be conscious of their responsibility and follow related information, such as the Norwegian Environment Agency's webpage on local air quality, which is recommended for kindergartens according to Geir Tore Aamdal. Cathrine Telstø Pedersen believes that kindergartens' trust in the authorities and their interest in air quality may be limited by the numerous other factors they need to consider in a kindergarten.

All surveyed early childhood teachers expressed that they knew about the municipality's alert system; however, four of the informants had limited knowledge about air quality in Stavanger. Aamdal said that Stavanger municipality has worked to improve air quality since 1998, and the related public service was established in 2016. He explained that the webpages are continually improved to become more user-friendly. Also, Stavanger Municipality has implemented measures such as road cleaning, depositing old wood stoves, and promoting public transportation to prevent air quality levels that would require alert systems to actively inform vulnerable citizens, similar to those in other European cities (Stauffer, 2017).

Only one of the informants answered that the kindergarten had guidelines related to air quality; the others were unsure. Four informants mentioned routines for dealing with indoor air quality. Telstø Pedersen suggested that kindergartens give priority to ensuring children's well-being inside. This may imply uncertainties between handling serious outdoor pollution with an alert from the municipality medical officer and low-risk situations that could still be significant for individual vulnerable children.

Only two informants reported knowing that air pollution affects children more than an average adult. Only one informant knew that NO_2 and particulate matter ($PM_{2.5}$ and PM_{10}) are harmful to individuals with airway-related health conditions (cf. Stavanger Municipality, 2018; WHO, 2021); the others indicated that they knew either little or nothing about these substances. Telstø Pedersen underlined that kindergartens take responsibility for risk evaluation plans for children with special health conditions and may become aware of air quality in this context. However, since PM threatens health at relatively low concentrations (WHO, 2021), increasing awareness of air pollution among kindergarten staff should be set in focus. Aamdal reported that very few kindergartens have contacted him with concerns (approximately three during the last 5 years), and these inquiries are likely related to children with special needs or parents' concerns. Aamdal is careful not to alarm the kindergartens unnecessarily or exaggerate the risks.

Three informants with medium to high interest in air quality reported following available information, although mainly in specific situations. One early childhood teacher mentioned the winter season, ensuring that children do not sleep outside during particularly cold weather. In Norway, the youngest children in kindergarten often have their daytime naps outdoors in their prams. Another informant avoids going outside during rush hours, especially on cold days in winter. These clear, cold winter days, especially those with little precipitation, can result in increased air pollution, mainly due to hazardous gases (NO_x) and airborne particulate matter from road wear and combustion (Stavanger Municipality, 2018). Aamdal pointed out that only a few cold winter days may pose challenges regarding air pollution, with private wood-burning heaters being the main source of harmful combustion in Stavanger, alongside road traffic, especially along the E39 highway. Most days of the year are recommended for outdoor activities. One informant explained that an alert would be communicated to parents and staff, and this might lead to measures such as reducing outdoor time for the children. According to Aamdal, the ultimate goal must be to prevent a worst-case scenario where kindergarten children should be held inside in periods of increased air pollution.

Two informants reported that parents have generally become more interested in environmental issues and sustainability, whereas five informants evaluated parents' interest in air quality and pollution as low. According to the informants in this study, less than half of the children were transported to and from kindergarten by car, followed by bicycle and walking, scooter, and public transport. This estimated modal split seems to reflect more environmentally conscious travel behaviour compared to the higher reliance on cars for such escort trips in Stavanger, as noted by Hernández-Palacio and Kesarovski (2024). Aamdal mentioned measures taken to increase public transport related to the national goal of zero growth in single-passenger traffic by 2030 (Stavanger Municipality, 2018); see also Joudavi and Tarigan, chapter 13, on cycling.

It is possible that the informants did not feel that the questionnaire called upon their general knowledge about local topics, or they may not have connected 'air pollution in the neighbourhood' with Stavanger Municipality's goal of becoming climate neutral by 2030 (European Commission, 2024; Gagliore, 2023; Stavanger Municipality, 2018). Kopnina (2017) underlined the importance of citizen participation and the development of personal and collective competencies for promoting environmental protection, human health, and the integrity of the closely interrelated ecological systems in the urban community (cf. United Nations, 2015). Two of the informants, both with medium to high interest, reported that they try to stay updated and follow recommendations from the municipality; others were uncertain about how they relate to air quality in their work. Duran (2021) underlined that individual awareness of the environment and conscious behaviour towards environmental problems are crucial first steps in preventing pollution; social awareness is also essential for reducing or preventing industrial air pollution.

None of the early childhood teachers typically discuss air quality with the children. One informant noted that they communicate acute situations, such as a fire or extreme industrial odours, which can be exciting and memorable for children. These events may serve as a good starting point for communicating interrelations in their urban environment (Sageidet et al., 2019). Telstø Pedersen envisioned playful learning inquiries with children as young as 5 years to introduce basic knowledge related to air quality, despite the theme's complexity. Literature also encourages early childhood teachers to address even complex interrelationships in simplified ways, as young children develop their knowledge gradually (Duran, 2021; Kos et al., 2016; Sageidet et al., 2019; Vygotzky, 1986). For example, children can be assisted in growing their awareness of the differences between various areas in their urban neighbourhoods, such as parks and roads, including the associated air quality. Guiding children through local roads and city areas can enhance their spatial abilities, visual reasoning, digital competencies, understanding of interrelationships, and critical thinking. As a former early childhood teacher, Telstø Pedersen would include visual methods (cf. Duran, 2021), such as using a picture showing healthy plants and humans thriving in fresh air in a park, contrasted with a greyish picture showing dirty roads without vegetation. Children may also understand that maps, like the one shown in Figure 14.1, may illustrate that the air in areas with dense buildings and roads is more likely to be polluted than in green areas (Borck & Schrauth, 2021; Duran, 2021; Liang & Gong, 2020; Sageidet et al., 2019; Zhivkov & Kesarovski, 2024).

Sustainability is a core value in the Norwegian kindergarten curriculum, even though air quality is not specifically addressed. The SDGs aim to empower children and youth (United Nations, 2015, Points 23 and 25), recognizing them as crucial groups for education in sustainability and responsible citizenship (Bell, 2016; Sageidet & Heggen, 2021). Kopnina (2017) highlighted interrelationships between environmental challenges and climate change, air pollution, children's health, and technology, underlining the role of education in this connection. (Early childhood) teachers can help children develop initial competencies to prioritize parks and green areas with clean air and rich biodiversity for playing and learning. Additionally, children and youth can learn why cycling and walking are healthier alternatives to travelling by car. Telstø Pedersen is confident that young children can even motivate their parents to choose air quality–friendly options, such as green spaces, over areas near heavy vehicle traffic in their everyday lives (cf. Kopnina, 2017; Sageidet et al., 2019).

Telstø Pedersen suggested planning target activities for and with children together with a professional. Aamdal agreed that knowledge is always useful, and even young children can benefit from small stepwise introduction to air quality issues, helping to establish good habits. He can imagine explaining to children the interiors of the small houses equipped with air quality monitors. Such visits, accompanied by simple explanations, can provide even very young children with a foundational understanding of connections between their local neighbourhood, technology, and environmental parameters like air quality/pollution. Through these experiences, (early childhood) teachers can promote the development of children's and youths' competencies and agency related to their local and global community. Children seek meaningful relationships and strive to connect to new information to their prior knowledge (Vygotzky, 1986). Such opportunities for children highlight the connection between sustainability and technology, fostering children's and youths' understanding, participation, and agency in smart city initiatives, ultimately preparing them to become active and informed citizens in a sustainable society (Montalvan Castilla & Müller, 2023; Sageidet & Heggen, 2021; United Nations, 2015).

Given the increasingly complex systems in cities, further research is needed on how to promote children's initial understanding of intertwined and holistic aspects of science, technology, engineering, and mathematics (STEM) in community and urban landscape contexts. However, it is important to recognize that exposure to air pollutants is beyond the control of kindergarten staff and individuals while addressing this issue requires actions by public authorities (WHO, 2021). Watts et al. (2015) underlined that a well-planned and improved public transport system, along with good walking and cycling infrastructure, can encourage mobility without relying on cars. This approach can provide alternative choices for dropping off and picking up kindergarten children, thereby reducing air pollution and greenhouse gas emissions.

Conclusions and implications

Within the Stavanger smart sustainable city approach, there is some untapped potential to disseminate information on reduced air quality at certain times, mainly caused by heavy traffic and private wood heating. Enhancing this information flow is essential for preventing negative health effects of air pollution on children in cities. Increasing awareness among citizens, especially those responsible for children, is also crucial for engaging them in initial communication of these issues with children and youth.

(Early childhood) teachers can play a significant role in promoting children's initial understanding of their complex urban environment; for example, by focusing on air quality. This includes initial introductions to air's physical and environmental properties, the concept of air pollution, and the technology of air pollution monitors in their neighbourhoods. Furthermore, children should be introduced to interrelationships related to air quality, such as the impact of car traffic versus the benefits of green spaces and parks. Such knowledge can empower children and youth to make more sustainable choices and encourage them to participate in decision-making processes within their communities.

Note

1 Yr.no is a popular weather forecasting service in Norway, offering real-time weather data, hourly and long-term forecasts, radar images, and alerts about severe weather conditions. It is a collaboration between the Norwegian Broadcasting Corporation and the Norwegian Meteorological Institute.

References

Baklanov, A., Molina, L.T., & Gauss, M. (2016). Megacities, air quality and climate. *Atmospheric Environment*, 126, 235–249. doi:10.1016/j.atmosenv.2015.11.059

Beckett, C., & Clegg, S. (2007). Qualitative data from a postal questionnaire: Questioning the presumption of the value of presence. *International Journal of Social Research Methodology*, 10(4), 307–317. doi:10.1080/13645570701401214

Bell, D.V.J. (2016). Twenty-first century education: Transformative education for sustainability and responsible citizenship. *Journal of Teacher Education for Sustainability*, 18(1), 48–56. doi:10.1515/jtes-2016-0004

Bennett, W.D., Zeman, K.L., & Kim, C. (1996). Variability of fine particle deposition in healthy adults: Effect of age and gender. *American Journal of Respiratory and Critical Care Medicine*, 153, 1641–1647.

Bettiol, A., Gelain, E., Milanesio, E., Asta, F., & Rusconi, F. (2021). The first 1000 days of life: Traffic-related air pollution and development of wheezing and asthma in childhood. A systematic review of birth cohort studies. *Environmental Health*, 20(1), 46. doi:10.1186/s12940-021-00728-9

Borck, R., & Schrauth, P. (2021). Population density and urban air quality. *Regional Science and Urban Economics*, 86, 103596. doi:10.1016/j.regsciurbeco.2020.103596

Castell, N., Schneider, P., Grossberndt, S., Fredriksen, M.F., Sousa-Santos, G., Vogt, M., & Bartonova, A. (2018). Localized real time information on outdoor air quality at kindergartens in Oslo, Norway, using low-cost sensor nodes. *Environmental Research*, 165(2018), 410–419. https://www.sciencedirect.com/science/article/pii/S0013935117316584

The Constitution of the Kingdom of Norway. (2014). *C. Rights of Citizens and the Legislative Power*. Article 112 (LOV-2014-05-13-112). Lovdata.

Davila, S., Ilić, J.P., & Bešlić, I. (2015). Real-time dissemination of air quality information using data streams and Web technologies: Linking air quality to health risks in urban areas. *Arh Hig Rada Toksikol*, 66(2), 171–180. doi:10.1515/aiht-2015-66-2633

Derry, S., Schunn, C.D., & Gernsbacher, M.A. (Eds.). (2014). *Interdisciplinary collaboration: An emerging cognitive science*. Psychology Press.

DOGA, Norwegian Smart City Network & Nordic Edge. (2019). *Roadmap for smart and sustainable cities and communities in Norway. A guide for local and regional authorities developed by Design and Architecture Norway (DOGA), the Norwegian Smart City Network and Nordic Edge*. https://nordicedge.org/projects/national-smart-city-roadmap/

Duran, M. (2021). Perception of preschool children about environmental pollution. *Journal of Education in Science, Environment and Health*, 7(3), 200–219. doi:10.21891/jeseh.733800

European Commission. (2024). *EU missions, 100 climate-neutral and smart cities – Cities on a journey to climate neutrality*. Publications Office of the European Union, Directorate-General for Research and Innovation. https://data.europe.eu/doi/10.2777/169604

Gaglione, F. (2023). Policies and practices of transition towards climate-neutral and smart cities. *TeMA - Journal of Land Use, Mobility and Environment*, 16(1), 227–231. doi:10.6093/1970-9870/9822

Garcia, E., Rice, M.B., & Gold, D.R. (2021). Air pollution and lung function in children. *Journal of Allergy and Clinical Immunology* 148(1), 1–14. doi:10.1016/j.jaci.2021.05.006

Greenwood, N.N., & Earnshaw, A. (1997). *Chemistry of the elements* (2nd ed.). Butterworth-Heinemann.

Heggen, P.M., Sageidet, B.M., Goga, N., Grindheim, L.T., Bergan, V., Utsi, T.A., Wallem Krempig, I., & Lyngård, A.M. (2019). Children as eco-citizens? *Nordic Studies in Science Education*, 15(4), 387–402. doi:10.5617/nordina.6186

Hernández-Palacio, F., & Kesarovski, T. (2024). Urban density and accessibility: A methodological approach. *Nordic Journal of Architectural Research*, 36(1), 89–116. https://hdl.handle.net/11250/3134394

Holst, G.J., Pedersen, C.B., Thygesen, M., Brandt, J., Geels, C., Bønløkke, J.H., & Sigsgaard, T. (2020). Air pollution and family related determinants of asthma onset and persistent wheezing in children: Nationwide case-control study. *BMJ*, 370, 2791. doi:10.1136/bmj.m2791

IQAir. (2020). *World air quality report: Region & city PM2.5 ranking*. https://www.greenpeace.org/static/planet4-romania-stateless/2021/03/d8050eab-2020-world_air_quality_report.pdf

Jupskås, S.H., & Römcke, F. (2002, April 11). Naturen er til å hoppe i [Nature is to jump in]. *Stavanger Aftenblad*. https://www.aftenbladet.no/lokalt/i/KW7y4/naturen-er-til-aa-hoppe-

Kopnina, H. (2017). Vehicular air pollution and asthma: Implications for education for health and environmental sustainability. *Local Environment*, 22(1), 38–48. doi:10.1080/13549839.2016.1154519

Kos, M., Jerman, J., Anzlovar, U., & Torkar, G. (2016). Preschool children's understanding of pro-environmental behaviour: Is it too hard for them? *International Journal of Environmental and Science Education*, 11(12), 5554–5571. https://files.eric.ed.gov/fulltext/EJ1115643.pdf

Lee, W.O., & Fouts, J. (Eds.). (2005). *Education for social citizenship. Perceptions from teachers in the USA, Australia, England, Russia and China*. University Press.

Liang, L., & Gong, P. (2020). Urban and air pollution: A multi-city study of long-term effects of urban landscape patterns on air quality trends. *Scientific Reports*, 10(1), 18618. doi:10.1038/s41598-020-74524-9

Liu, M.-M., Wang, D., Zhao, Y., Liu, Y.-Q., Huang, M.-M., Liu, Y., Sun, J., Ren, W.-H., Zhao, Y.-D., He, Q.-C., & Dong, G.-H. (2013). Effects of outdoor and indoor air pollution on respiratory health of Chinese children from 50 kindergartens. *Journal of Epidemiology*, 23(4), 280–287. doi:10.2188/jea.JE20120175

Maione, M., Mocca, E., Eisfeld, K., Kazepov, Y., & Fuzzi, S. (2021). Public perception of air pollution sources across Europe. *Ambio*, 50, 1150–1158. doi:10.1007/s13280-020-01450-5

Mayring, P. (2014). *Qualitative content analysis: Theoretical foundation, basic procedures, and software solutions*. Klagenfurth. http://nbn-resolving.de/urn:nbn:de:0168-ssoar-395173

Montalvan Castilla, J.E., & Müller, A. (2023). A smart city for all citizens: An exploration of children's participation in Norway's smartest city. *International Planning Studies*, 29(1), 19–33. doi:10.1080/13563475.2023.2259110

Norwegian Institute of Public Health. (2018). *Health status in Norway 2018*. Public health report – Short version. https://www.fhi.no/contentassets/d021a759c5ed48a e85fffc94e35785cf/health_status_in_norway_2018.pdf

Rehwagen, M., Schlink, U., Herbarth, O., & Fritz, G.J. (1999). Pollution profiles at different kindergarten sites in Leipzig, Germany. *Environmental Toxicology*, 14, 321–327. doi:10.1002/(SICI)1522–7278(199907)14:3<321:AID-TOX5>3.0.CO;2-L

Rodriguez, C., Tonkin, R., Heyworth, J., Kusel, M., De Klerk, N., Sly, P.D., Franklin, P., Runnion, T., Blockley, A., Landau, L., & Hinwood, A.L. (2007). The relationship between outdoor air quality and respiratory symptoms in young children. *International Journal of Environmental Health Research*, 17(5), 351–360. doi:10.1080/09603120701628669

Sageidet, B.M., Almeida, S.C., & Dunkley, R. (2018). Children's access to urban gardens in Norway, India and the United Kingdom. *International Journal of Environmental and Science Education*, 13(5), 467–480. http://www.ijese.net/makale/2058.html

Sageidet, B.M., Christensen, M., & Davis, J.M. (2019). Norwegian kindergarten children's understandings of sustainability related issues in comparison to their peers' understandings in Australia. *International Journal of Environmental and Science Education*, 14(4), 191–205. http://www.ijese.net/makale/2115.html

Sageidet, B.M., & Heggen, M.P. (2021). Global citizenship and the Sustainable Development Goals. In W. Leal Filho, P. Gökçin Özuyar, A.M. Azul, L. Brandli, U. Azeiteiro, & T. Wall (Eds.), *Reduced inequalities. Encyclopedia of the UN Sustainable Development Goals* (pp. 1–11). Springer. doi:10.1007/978-3-319-71060-0_45-1

Schneider, P., Vogt, M., Haugen, R., Hassani, A., Castell, N., Dauge, F.R., & Bartonova, A. (2023). Deployment and evaluation of a network of open low-cost air quality sensor systems. *Atmosphere*, 2023(14), 540. doi:10.3390/atmos14030540

Squizzato, S., Cazzaro, M., Innocente, E., Visin, F., Hopke, P.K., & Rampazzo, G. (2017). Urban air quality in a mid-size city – PM2.5 composition, sources and identification of impact areas: From local to long range contributions. *Atmospheric Research*, 186, 51–62. doi:10.1016/j.atmosres.2016.11.011

Stauffer, A. (2017). *Toolkit: Communicating on air quality and health. Urban Agenda for the EU*. https://futurium.ec.europa.eu/en/urban-agenda/air-quality/library/comm unication-toolkit-communicating-air-quality-and-health

Stavanger Municipality. (2018). *Climate and environmental plan 2018–2030*. https:// www.stavanger.kommune.no/finn/?q=climate%20and%20environmental%20plan

United Nations. (2015). *Transforming our world: The 2030 agenda for sustainable development*. https://sdgs.un.org/2030agenda

Vygotzky, L.S. (1986). *Thought and language.* Harvard University Press.

Watts, N., Neil Agder, W., Agnolucci, P., Blackstock, J., Byass, P., & Cai, W. (2015). Health and climate change: Policy responses to protect public health. *Lancet,* 386, 1861–1914.

World Health Organization. (2006). *Air quality guidelines: Global update 2005: Particulate matter, ozone, nitrogen dioxide, and sulfur dioxide.* https://www.who.int/publications/i/item/WHO-SDE-PHE-OEH-06.02

World Health Organization. (2021). *Review of evidence on health aspects of air pollution: REVIHAAP project: Technical report.* https://apps.who.int/iris/handle/10665/341712

Wu, Y., Zhang, L., Wang, J., & Mou, Y. (2021). Communicating air quality index information: Effects of different styles on individuals' risk perception and precaution intention. *International Journal of Environmental Research and Public Health,* 18(19), 10542. doi:10.3390/ijerph181910542

Yu, H., & Zhang, H. (2023). Impact of ambient air pollution on physical activity and sedentary behaviour in children. *BMC Public Health,* 23, 257. doi:10.1186/s12889-023-15269-8

Zhivkov, P., & Kesarovski, T. (2024). Dynamic relationship between population densities and air quality in the four largest Norwegian cities. In *Annals of Computer Science and Information Systems: Vol. 39. Proceedings of the 19th Conference on Computer Science and Intelligence Systems (FedCSIS)* (pp. 713–718). https://annals-csis.org/proceedings/2024/pliks/fedcsis.pdf

Zhong, J., Hodgson, J.R., Bloss, W.J., & Shi, Z. (2023). Impacts of net zero policies on air quality in a metropolitan area of the United Kingdom: Towards world health organization air quality guidelines. *Environmental Research,* 236(1), 116704. doi:10.101016/j.envres.2023.116704

15 Making art smart

Kristiane M.F. Lindland and Fredrik A.F. Matre

Introduction

Stavanger City Council has defined Urban Art as one of the five pillars constituting their smart city concept (Stavanger Municipality, 2016). This chapter illustrates and discusses how art can become a performative element in the development of the smart city. The municipal decision to draft a smart city strategy was in part a response to an economic recession starting in 2014 as a result of falling oil prices. As Norway's 'oil capital', the downturn of global petroleum markets had a marked negative impact on a sizable proportion of employers and employees within the municipality. This economic hardship drew greater attention towards the need for plans facilitating an economic future less reliant on petroleum extraction, with calls to pivot towards renewable energy and digital industries. Drawing on successful past experiences of championing urban art, the municipality included it as a pillar in their smart city strategy: the ambiguity of its definition, however, created an opening for interpretation and accommodation by future collaborators.

Art has, throughout the ages, inspired and informed notions of the possible. It is the power of art to speak to innate human emotions that makes one seek out galleries displaying great swathes of creative endeavours. Art, beyond mere aesthetic, can engender critical discussions about societal problematics. Stavanger has some historic form with promoting art in the public realm, hosting the annual street art festival NuArt since 2001 (NuArt, 2024) and being selected as European Capital of Culture (alongside Liverpool) in 2008 (see Bergsgard & Vassenden, 2011). This historical precedent may explain why Stavanger City Council chose to specify 'urban art' rather than 'art' or 'smart art' as part of their smart city strategy. While art in the public realm by default engages with passers-by who look up, conventional art institutions like museums must entice people through their doors. To be successful, art institutions need to both be able to present artworks of general interest as well as be able to tell a story about relevance that engages the public and turns citizens into visitors. Hence, they continually strive to present their pieces in ways that feel relevant to contemporary societal interests and debates, often by recontextualizing the artwork.

Physical objects can serve as tools for novel imaginations through alternative interpretations of reality, informed by re-interpretations of the past. Dewey's

DOI: 10.4324/9781003498650-19

(1938), Dewey and Bentley's (1949), Follett's (1924), and Mead's (1932, 1934) understandings of meaning, identity, and temporality provide opportunities for exploring how an exhibition of historical art can create movement in meaning, imaginations, and selves in our present. Based on interviews, field conversations, and participatory observations related to an exhibition of painter Kitty Kielland's work, this chapter explores how smart art can be used strategically for new ways of achieving funding, new collaborations, and new ways of curating historical art. Following this, it explores how historical art can be recontextualized using innovative technology, can become a performative element in the development of the sustainable, smart city, beyond what was probably imagined by those developing the smart city strategy of Stavanger.

Art in smart city research

As a fairly new subject of academic study – Ingwersen and Serrano-López's (2018, p. 1205) find that 'the first mention of the concept "smart city(ies)" in publication titles takes place in 1999' – one should expect knowledge gaps in smart city research that need interrogation. While the concept of smart city remains, at least for now, an essentially contested concept, this plurality of conceptual views also lends itself to great cross-disciplinary potential for debate.

Although literature on conceptualizations of the smart city abounds – including literature reviews (Anthopoulos, 2015; Zhao et al., 2021), state-of-the-field surveys (Kirimtat et al., 2020; Neirotti et al., 2014), and proposals for a unified conception of the smart city (Ramaprasad et al., 2017; Zubizarreta et al., 2015) – less attention has been paid to how smart art fits into the concept. Specifically, scholars (such as Krivý, 2018) rightly point out the need to interrogate smart city as more than technocratic ordering. An overemphasis on technology can be a hindrance in smart city research, as Angelidou (2014, p. S3) succinctly emphasized:

> The tendency to believe that innovative technological instrumentation automatically transforms a city into a 'smart' one, and a biased use of the buzzword 'smart' in fragmented or superficial ways, actually hinder[s] the clarification of the subject even further.

While the use of technology is part and parcel in almost all explicit applications of a smart city strategy, we follow from Komninos et al. (2016) in looking at smart city (and more specifically smart art) from an ontological standpoint first and a technological one second, while deviating from Komninos et al. in applying a philosophical/ sociological rather than information technology understanding of ontology.

Hatem's case study of Milton Keynes' incorporation of art in a smart city strategy is a welcome start in elucidating how 'the role of art' can go 'beyond that of decoration' and can in fact 'help facilitate cities' development and production' (Hatem, 2023, p. 291). The case echoes similar (yet distinct) findings of Eynaud et al. (2018) showing how participatory art in the urban realm can be a social practice of communing, asserting what Henri Lefebvre called the 'right to the city'. Anderson (2017) described a case from Manchester where performative art was used to better engage citizens in smart city initiatives, helping participants to learn about the rationale behind new developments and also voice their concerns and ideas. Littwin and Stock (2020), furthermore, have found 'strong hints' that interactive contemporary art installations in the urban public realm 'may symbolize the city's smartness' (p. 20). While these smart city–focused contributions are merited, they do not satisfy our curiosity regarding what makes art smart: beyond use of new technologies, art must have a further quality to be smart.

The case presented in this chapter stems from a particular application of smart city thinking, originating in a strategy drawn up by municipal authorities but further developed by practitioners. On December 12, 2016, Stavanger City Council adopted a roadmap for smart city developments in the municipality. It identified three prerequisites for a project to be considered a smart city project: technology ('to simplify and improve'), cooperation ('across local authorities, industry and commerce, organisations and academia'), and citizen involvement ('based on the citizens' and users' needs'; Stavanger Municipality, 2016). The roadmap further identified five priority areas, one of which is 'Urban Art'.[1]

Montalvan Castilla and Müller (2024) found that, partially in response to criticism of smart city projects being fixated on technological quick-fixes and technocratic top-down micro-governance, many European smart city initiatives have correspondingly 'shifted towards so-called people-centred smart cities emphasizing the importance of citizen participation' (p. 19). We agree with Montalvan Castilla and Müller that Stavanger's smart city approach is an example of this 'participatory turn' in smart city applications, generally observed in European cities such as those involved in New European Bauhaus projects.[2] The city council has continued placing steady emphasis on cross-sectoral collaboration in realizing smart city projects in the greater urban environment, such as the Nordic Edge Expo hosted in Stavanger. Nordic Edge is a non-profit that serves as a collaborative forum on urban innovation (Nordic Edge, 2024). Their 2017 expo was at the time billed as the largest smart city event in the Nordics, and it included a 'safari' through the city with a stop at Stavanger Art Museum (a stone's throw away from the conference centre hosting the expo) to experience their new exhibition *Fri Luft* showcasing a comprehensive collection of works by local artist Kitty Kielland (1843–1914).

A pragmatist approach to understanding art as a dialogical process of becoming

The empirical case examined here can be understood as an instance of strategy execution. A strategy can be understood as a plan, a plot, or a 'roadmap' (Karlsen, 2016, p. 15; Mintzberg, 1987). These understandings are all so-called substantive understandings of the term. In this chapter we take a very different approach to strategy, seeing it as a dynamically developing process that is both situated and relational. This means that a strategy is never 'finished'; its meaning is continually developing through how people use the strategy for different purposes and thus add and reconsider different meanings to it. By calling it 'situated', we mean that it is continuously developed and realized in a specific context (physically, temporally, and socially) and thus related to specific events. Finally, we refer to it as relational since it is not defined by anyone alone but continuously developed through gestural conversations between different individuals and groups who are both forming and being formed by the development of events (Mead, 1934).

Seeing strategy as a process can be done in different ways. One direction that has gained more attention in recent years is what has been labelled as strategy-as-practice (see, for example, Jarzabkowski & Spee, 2009; Jarzabkowski & Whittington, 2008). Although we align with this thinking to some extent, we take what Cloutier and Langley (2020) call a radical process perspective to strategy and ground our understanding in American Pragmatism.

Four elements are central to this understanding: transactionality, temporality, habit, and inquiry. *Transactionality* is the idea that action is something not solely happening between actors but simultaneously happening through actors. In other words, actors are influenced by the emergence of events, but the events are also continuously influenced by the developing actors. This understanding is also connected to a temporal dimension, as we do not see situations or actors and events as static. They are continuously developing, co-constituted through the situation. Our understanding of *temporality* follows Mead (1932): reality only exists in the present, while our understanding of the past and expectations for the future are continuously re-constructed through reality as we experience it in the present. Mead (1932, 1934) proposed that meaning developed through gestural conversations of gesture–response, where a gesture (an event, utterance, or body language) evokes some response in those related to it. This response is then a gesture for the first actor to respond to. However, the response is based not solely on the interpretation of the gesture itself but on the imagination of how the other will respond to the imagined and prospective gesture. Therefore, gestural conversations have both internal and external dimensions, and the positions people take are continuously adapted to expectations of the developing situation. Through numerous transactions in similar situations, we develop *habits* of how to respond to specific events. These are not routines but responses that have proved to work well in similar situations. As such, they can also be adapted and changed.

Sometimes we receive a completely different response than anticipated. This leads to a doubtful present, which forces us to reconsider our understanding of the situation and consequently reconsider our understanding of past and possible future actions: what Dewey (1938) and Dewey and Bentley (1949) called *inquiries*. We search for alternative understandings that can make better sense of the present moment. An alternative interpretation of the situation can make the situation temporarily stable and enable us to act upon what we imagine to be the situation. In these re-considerations we also change our perceptions of one another and who we become in relation to others.

Through illustrative examples that emerged from the Kitty Kielland exhibition, we will give insight into how the smart city strategy of Stavanger City Council was responded to by the museum and how the development of events possibly led to new understandings of what the museum and the visitors could become and what futures the historical art could help us imagine.

Methods

The empirical basis for this chapter comes from a preliminary, qualitative study of innovation in the museum sector. The preliminary study was initiated by the first author's initiative and a conference call on the influence of cultural cities on social development. As such, one of the aims of the study was to examine how Stavanger's year as European Capital of Culture in 2008 has influenced the development of innovation in art dissemination at the Stavanger Art Museum.

The preliminary study consisted of four in-depth interviews with key persons in the development of the *Fri Luft* exhibition, three of whom were from the museum and one from a communications company working as a technical partner in the execution of the exhibition. The first author followed two tours of the art exhibition: one tour that was part of the innovation conference Nordic Edge's program and another tour with a primary school class. The second author experienced the exhibition and the responses to it from another angle, employed as a receptionist and café host at the museum.

The transactional situations described in the interviews and observed during the museum activities are here interpreted and discussed with emphasis on how they appear to create *movement in meaning* (Carroll & Simpson, 2012): in both who we become in the meeting with the historical art and how the art helps us see our own time and environment in a different way. Obtaining insights from internal parties (obtained by the second author) and external observers (obtained by the first author), we have gathered a greater breadth of material to fully appreciate the dynamics of the case, which further helps question initial assumptions and more faithfully represent the views of informants as abductive inquiry (Lorino, 2014).

As pragmatists we are not seeking definite truths, nor do we even think there can be a definite and united understanding of social situations. All situations will be understood more or less differently by the actors taking part in the situation. Hence, there cannot be one definitive truth regarding how to understand a situation. However, through being socialized into different contexts and situations, we have developed more or less shared understandings of how to interpret various gestures (Mead, 1934). Consequently, we become able to respond meaningfully to gestures and to have relevant expectations for how others will respond to the same gestures in the situation (Mead, 1934). What we hereby suggest are plausible hypotheses about what happened, using these to further explore the possible roles of art in developing sustainable urban futures (Lorino, 2014; Wegener & Lorino, 2020).

Kitty Kielland then and now: curatorial innovation

Stavanger Art Museum stated that they had long wished to organize an exhibition of the works of Christine 'Kitty' Lange Kielland (1843–1914) A celebrated artist even in her own time (Thiis, 1907, cited in Gudmundson, 2017) she remains one of the most prominent female Norwegian painters in art history (Gudmundson, 2017; Ueland et al., 2017). Born into the wealthy and influential Kielland family, she enjoyed considerable freedom to pursue her artistic passion, much like her celebrated writer-brother Alexander Kielland. Having studied art in Germany and France, including exhibiting works at the Paris Salon, Kielland was an active social commentator (Åsebø & Utne, 2017; Kielland, 1886). Her remarks on female emancipation are increasingly seen as equally influential to her paintings (Sjåstad, 2020), which followed the prominent naturalistic-realistic style of the day and later contributed to the development of Nordic Neo-Romanticism.

Kielland's work had not been shown as part of a stand-alone exhibition in Norway before. Between June and October 2017, Stavanger Art Museum exhibited a vast collection of Kielland's works in an exhibition entitled *Fri Luft* ('Open Air', from the French *plein air*). Her works were exhibited alongside select pieces by her contemporaries, including celebrated artists Hans Gude (1825–1903) and Harriet Backer (1845–1932). There was great interest from the general public: Kielland is, after all, a familiar name in her hometown of Stavanger, and this exhibition was one of the first to bring together such an extensive collection of her works, with many depicting local scenes. She is most recognized for her paintings depicting peat bogs in the nearby Jæren countryside; these paintings, in particular, were what locals came to see.

Alongside the familiar accoutrement of wall text conveying biographical and contextual information about the art and artists, there was also the opportunity to put on a fairly large augmented reality (AR)/mixed reality (MR) headset. The museum had contracted a local media company, Hey-Ho Let's Go, to create tie-in applications for the exhibition. Two of them, Kitty Filter and Kitty Motiv were mobile apps available on iOS and Android devices. Kitty Filter let users apply filters to their photographs that lent them the appearance of oil paintings in the naturalistic-realistic style, while Kitty Motiv was meant to challenge the public into identifying exactly where Kielland painted her motifs of Jæren in the open air.[3]

The final application, Holo Kitty, was developed for use with the Microsoft HoloLens AR/MR headset. While in the gallery in front of one of Kielland's paintings, one could don the headset and be placed in front of an easel. Surrounding one would be the familiar environment Kielland would have experienced as she painted the picture, with butterflies flying by and birds singing in one's ear. Rather than a virtual reality experience where the outside world is no longer accessible to the participant, Holo Kitty lets the past and present intermingle. The tangible art object on the wall is reimagined into a co-constitutive piece of art expression as the observer becomes a participant and the object is metaphorically transported over a hundred years back to its creation (see Stavanger Kunstmuseum, 2017).

Framing 'old art' as smart: from strategy stipulation to implementation

As with much smart city research, it can be alluring to focus predominantly on the innovative technological aspects of art experiences. The use of AR/MR technology in presenting *Fri Luft* was, especially at the time, a novelty, in terms both of it being the first time the museum had used such technology as well as of it being the first interaction most audience members had with said technology. Yet, it was emphasized by informants organizing the exhibition that the technology was not employed for its own sake: rather, it was utilized for its ability to engage audiences in a unique way that would be beneficial to fulfilling their mission of informing and educating the public. Nevertheless, the technology is both a tool for creating new art experiences as well as something that can appeal to new groups of the art public.

To show different aspects of how the *Fri Luft* exhibition can be a form of Smart Art, we will first discuss how the municipality's stated smart city strategy was accommodated and responded to as a gestural invitation for collaboration. We then move on to seeing how the different objectives of stakeholders informed how the exhibition and the art objects were presented, recontextualizing the art experience to relate it to contemporary societal debates.

Stavanger Art Museum effectively seized on an opportunity for collaboration and funding to realize a long-standing objective. Wanting to put on a major exhibition of Kielland's work, they examined how they could pitch such an endeavour in a way that would suit the smart city objectives of the municipal authorities and related stakeholders. Seizing on the emphasis of urban art as a part of the smart city vision, the museum had to consider what curatorial choices would be needed, not to blindly conform to the confines of what is conventionally understood as 'urban art' but to pragmatically accommodate the smart city ethos into their exhibition.

Wishing to facilitate the creation of a greater body of research focused on Kitty Kielland and to do her substantive body of work justice in her hometown, the museum was in need of external collaborators and funding. Research is costly and time-consuming, and sourcing artworks on loan from other European museums as well as private collectors is logistically challenging and expensive (transportation and insurance are major expenditures). But the exhibition itself would also be a major expense, as explained by a curator:

> The physical curation of an exhibition is now very different than it was just a few years ago. There is more demand for making the whole physical framing around the art works as part of the art experience. For this exhibition we have used carpets, paint and even built a solitary room within the room for experiencing one of the paintings alone. We have rented designer furniture to 'furnish' the exhibition, which we would not have done just ten years ago.

The museum therefore framed the exhibition around smart city and smart art, effectively communicating their relevance to municipal priorities. They recruited relevant stakeholders, such as local media company Hey-Ho Let's Go, which was commissioned to create three tie-in applications for the exhibition that integrated augmented reality and interactive technology with the art experience. Furthermore, they partnered with the Nordic Edge conference to be included as one of their 'safaris' exploring urban experimentation emblematic of the city's smart city efforts, thereby legitimizing the exhibition as a bona fide smart city project. In exploring what smart art means, the museum implicated itself with the wider smart city environment.

A second major element to this case is recontextualization. One might think that historical works of art have little relevance for understanding the present. It may even seem counterintuitive to suggest that 'old art' can contribute much to contemporary debates on climate change, biodiversity loss, and energy challenges. Kielland's paintings of peat bogs, however, demonstrate how contemporary arable land in Jæren used to be under-cultivated peat landscapes. They thereby show how the natural landscape has been changed by humans to fulfil human needs. Modern environmental science now tells us that peat bogs are incredibly effective natural CO_2 sinks, and thus peat conservation is addressed in both global and national sustainability strategies (Landbruksdirektoratet, 2020; United Nations, 2015). One can therefore imagine how these cultivated landscapes have, through removing natural carbon mitigation measures, contributed to climate change and removed protections against its harms. As such, the exhibition contributes to contemporary environmental debates. These themes were capitalized upon by the exhibition organizers, who convened environmental conservation organizations, Nordic Edge, and the art museum in an attempt to mainstream socio-cultural discussions of the Anthropocene. Hence, the exhibition also led citizens to see and understand their surroundings with new eyes, which can increase attention to how human activities impact and are impacted by the local and global environment.

Discussion, contributions, and further suggestions to understanding smart art as a part of developing sustainable, smart cities

Embedded in an American Pragmatist understanding, especially Mead's (1932, 1934) understanding of meaning-making and meaning-makers as co-constituting one another through gestural conversations (see Lindland, chapter 5 for further elaboration on this), we see the art aspect of the smart city strategy, the Kielland exhibition, and the collaborations as a dialogue where one gesture evokes responses in others, which themselves function as gestures for others to respond to. In these gestural conversations, no party alone controls the development of events, and differences contribute to fostering novel imaginations about what could possibly be realized. On a broader point, we see how the presentation of the art objects having relevance to contemporary environmental debates indicated how smart art – much like smart city more broadly – can have an ethical dimension: the continuous striving of improving a current situation into a better one (Simpson & den Hond, 2022).

Smart art in the context of smart cities is broadly about how art can contribute to developing urban areas in inclusive ways. In urban development projects involving citizens, this often happens through planned and designed interaction with various technological as well as manual tools and activities. It can also happen through more bottom-up processes where citizens themselves 'impose' on an area through, for example, street art or by repurposing an area for a new type of use. We have here rather chosen to focus on a top-down approach. By drawing attention to how the meaning of a concept as well as imagined futures (regarding environmental stewardship) can change, one appreciates how historical art objects can contribute to fostering novel insights. Historical works of art are witnesses to our past and can tell of anthropogenic changes to both local landscapes and global climatic conditions.

Our research has drawn attention to how urban strategies for development, such as smart city strategies, create possibilities for new collaborations across institutional boundaries and interests. Motivation for doing so can, however, vary. In this case, we have focused on how the framing of smart city can evoke re-framings of historical art and thereby provide possibilities for new actuality, new sources of funding, and new forms of art curation.

For practitioners working with artistic methods as a tool for urban development and inclusion, it is thus important to pay attention in designing smart art activities that can engage citizens and give 'them a sense of involvement and uniqueness in a potentially homogenous and controlled environment' as they contribute to smart city initiatives (Shipman, 2019, p. 251). Instead of designing activities according to convention, we should direct our attention to how the way we conduct these projects and activities possibly also leads to re-production and re-confirmation of existing power imbalances. We therefore make the following suggestions for further research:

- How can collaborative activities of developing the smart city through art and artistic methods foster and cope with conflicting understandings to develop new understandings?
- How can we let citizens initiate and direct urban development using artistic methods through bottom-up processes when funding and legitimacy of urban collaborative projects usually place local government or private companies in the driver's seat?
- Discussing publicly accessible art through an American Pragmatist framework would usually draw extensively from Dewey and his *Art as Experience* (1934). We have not done so here, as the remit of this chapter does not include the public art experience in detail. This would, however, be a valuable effort and one that we hope to explore elsewhere.

Acknowledgements

The authors extend their sincere thanks to Stavanger Kunstmuseum, Museum Stavanger (MUST), and Hey-Ho Let's Go Ltd. for their great assistance in field-work during and after the *Fri Luft* exhibition. Gratitude is further extended to the anonymous reviewers who provided valuable commentary on prior drafts, the editorial team for excellent support, as well as Routledge for allowing for much-valued authorial freedom.

Declaration of interests

The authors declare no competing interests in preparing this manuscript. The research conducted received no external funding and was undertaken in a personal capacity. The views and opinions expressed are the authors' alone and do not reflect or represent their respective employers, nor should these organizations be understood to endorse or be affiliated with the arguments made in the work.

Notes

1 The other four priority areas are Health & Welfare; Education & Knowledge; Energy, Climate & Environment; and Governance & Democracy.
2 Note that Stavanger is involved in New European Bauhaus through the EU-funded NEB-STAR project.
3 One informant did note the superfluousness of this app, in that they had 'mapped most of the scenes already'.

References

Anderson, A. (2017, June 30). The art of smart cities. *International Arts Manager*. https://internationalartsmanager.com/the-art-of-smart-cities/

Angelidou, M. (2014). Smart city policies: A spatial approach. *Cities*, 41(Suppl. 1), S3–S11. doi:10.1016/j.cities.2014.06.007

Anthopoulos, L.G. (2015). Understanding the smart city domain: A literature review. In M.P. Rodríguez-Bolívar (Ed.), *Transforming city governments for successful smart cities* (pp. 9–21). Springer. doi:10.1007/978-3-319-03167-5

Åsebø, S., & Utne, J.M. (2017). Kitty L. Kiellands stemmer. In I.M.L. Gudmundson (Ed.), *Fri luft* (pp. 79–91). Uten Tittel.

Bergsgard, N.A., & Vassenden, A. (2011). The legacy of Stavanger as capital of culture in Europe 2008: Watershed or puff of wind? *International Journal of Cultural Policy*, 17(3), 301–320. doi:10.1080/10286632.2010.493214

Carroll, B., & Simpson, B. (2012). Capturing sociality in the movement between frames: An illustration from leadership development. *Human Relations*, 65(10), 1283–1309. doi:10.1177/0018726712451185

Cloutier, C., & Langley, A. (2020). What makes a process theoretical contribution? *Organization Theory*, 1(4), 1–32. doi:10.1177/2631787720902473

Dewey, J. (1934). *Art as experience*. Minton, Balch & Co.

Dewey, J. (1938). *Logic: The theory of inquiry*. Henry Holt and Company.

Dewey, J., & Bentley, A.F. (1949). *Knowing and the known*. Beacon Press.

Eynaud, P., Juan, M., & Mourey, D. (2018). Participatory art as a social practice of commoning to reinvent the right to the city. *Voluntas: International Journal of Voluntary and Nonprofit Organizations*, 29(4), 621–636. doi:10.1007/S11266-018-0006-y

Follett, M.P. (1924). *Creative experience*. Longmans, Green and Company.

Gudmundson, I.M.L. (2017). Introduksjon. In I.M.L. Gudmundson (Ed.), *Fri luft* (pp. 9–17). Uten Tittel.

Hatem, F.M. (2023). The role of art in smart cities research and making. *IET Smart Cities*, 5(4), 291–302. doi:10.1049/SMC2.12064

Ingwersen, P., & Serrano-López, A.E. (2018). Smart city research 1990–2016. *Scientometrics*, 117(2), 1205–1236. doi:10.1007/S11192-018-2901-9

Jarzabkowski, P., & Spee, A.P. (2009). Strategy-as-practice: A review and future directions for the field. *International Journal of Management Reviews*, 11(1), 69–95. doi:10.1111/J.1468-2370.2008.00250.X

Jarzabkowski, P., & Whittington, R. (2008). A strategy-as-practice approach to strategy research and education. *Journal of Management Inquiry*, 17(4), 282–286. doi:10.1177/1056492608318150

Karlsen, J.E. (2016). *Strategisk fremsynsledelse*. Fagbokforlaget as.

Kielland, K.L. (1886). *Kvindespørgsmaalet: Tilsvar til Hr. pastor J.M. Færden*. Thronsen & Co.

Kirimtat, A., Krejcar, O., Kertesz, A., & Tasgetiren, M.F. (2020). Future trends and current state of smart city concepts: A survey. *IEEE Access*, 8(1), 86448–86467. doi:10.1109/ACCESS.2020.2992441

Komninos, N., Bratsas, C., Kakderi, C., & Tsarchopoulos, P. (2016). Smart city ontologies: Improving the effectiveness of smart city applications. *Journal of Smart Cities*, 1(1), 31–46. doi:10.18063/JSC.2015.01.001

Krivý, M. (2018). Towards a critique of cybernetic urbanism: The smart city and the society of control. *Planning Theory*, 17(1), 8–30. doi:10.1177/1473095216645631

Landbruksdirektoratet. (2020). *Nasjonalt program for jordhelse*. https://www.landbruksdirektoratet.no/nb/nyhetsrom/rapporter/nasjonalt-program-for-jordhelse

Littwin, K., & Stock, W.G. (2020). Signaling smartness: Smart cities and digital art in public spaces. *Journal of Information Science Theory and Practice*, 8(1), 20–32. doi:10.1633/JISTaP.2020.8.1.2

Lorino, P. (2014). Charles Sanders Peirce (1839–1914). In J. Helin, T. Hernes, D. Hjorth, & R. Holt (Ed.), *The Oxford handbook of process philosophy and organization studies* (pp. 143–165). Oxford University Press. doi:10.1093/oxfordhb/9780199669356.001.0001

Mead, G.H. (1932). *The philosophy of the present*. Prometheus Books.

Mead, G.H. (1934). *Mind, self, and society*. University of Chicago Press.

Mintzberg, H. (1987). The strategy concept I: Five Ps for strategy. *California Management Review*, 30(1), 11–24. doi:10.2307/41165263

Montalvan Castilla, J.E., & Müller, A.R. (2024). A smart city for all citizens: An exploration of children's participation in Norway's smartest city. *International Planning Studies*, 29(1), 19–33. doi:10.1080/13563475.2023.2259110

Neirotti, P., De Marco, A., Cagliano, A.C., Mangano, G., & Scorrano, F. (2014). Current trends in smart city initiatives: Some stylised facts. *Cities*, 38(1), 25–36. doi:10.1016/J.CITIES.2013.12.010

Nordic Edge. (2024). *About us*. https://nordicedge.org/about-us/

NuArt. (2024). *About us*. https://archive.nuartfestival.no/about-us.html

Ramaprasad, A., Sánchez-Ortiz, A., & Syn, T. (2017). A unified definition of a smart city. *Electronic Government*, 10428, 13–24. doi:10.1007/978-3-319-64677-0_2

Shipman, H. (2019). Smart art for smart cities. In M. Mateev & P. Poutziouris (Eds.), *Creative business and social innovations for a sustainable future: Proceedings of the 1st American University in the Emirates International Research Conference – Dubai, UAE 2017* (pp. 251–253). Springer. doi:10.1007/978-3-030-01662-3_27

Simpson, B., & den Hond, F. (2022). The contemporary resonances of classical pragmatism for studying organization and organizing. *Organization Studies*, 43(1), 127–146. doi:10.1177/0170840621991689

Sjåstad, Ø. (2020). Kitty Kielland as a 'new woman'. *Scandinavian Studies*, 92(4), 492–520. doi:10.3368/sca.92.4.0492

Stavanger Kunstmuseum. (2017). *Fri Luft*. https://www.stavangerkunstmuseum.no/en/events/fri-luft

Stavanger Municipality. (2016, December 12). *Roadmap for the smart city Stavanger: Vision, goals and priority areas*. https://www.stavanger.kommune.no/siteassets/samfunnsutvikling/planer/engelske-planer/roadmap-smart-city-stavanger-2016.pdf

Ueland, H.B., Skeide, C., Pettersson, J.Å., Aavitsland, S., Ness, C., & Anker, N. (2017). Forord. In I.M.L. Gudmundson (Ed.), *Fri luft* (p. 5). Uten Tittel.

United Nations. (2015). *Transforming our world: The 2030 agenda for sustainable development.* https://sdgs.un.org/2030agenda

Wegener, F.E., & Lorino, P. (2020). Capturing the experience of living. In J. Reinecke, R. Suddaby, A. Langley, & H. Tsoukas (Eds.), *Time, temporality, and history in process organization studies* (pp. 138–168). Oxford University Press. doi:10.1093/oso/9780198870715.001.0001

Zhao, F., Fashola, O.I., Olarewaju, T.I., & Onwumere, I. (2021). Smart city research: A holistic and state-of-the-art literature review. *Cities,* 119(1), 103406. doi:10.1016/J.CITIES.2021.103406

Zubizarreta, I., Seravalli, A., & Arrizabalaga, S. (2015). Smart city concept: What it is and what it should be. *Journal of Urban Planning and Development,* 142(1), 04015005. doi:10.1061/(ASCE)UP.1943–5444.0000282

Part IV

Lessons learned

16 Lessons learned from a living smart city

Daniela Müller-Eie, Barbara Maria Sageidet, and Kristiane M.F. Lindland

And then Stavanger became smarter

Since 2019, the Research Network for Smart Sustainable Cities at the University of Stavanger (UiS), an inter-faculty initiative including researchers from the humanities, educational sciences, social sciences, and technical and natural sciences, has explored the narrative and development of Stavanger as a Nordic smart city over time and from an extraordinary range of disciplinary perspectives. The researchers and practitioners have had ample time to reflect on the smart city concept and observe smart city implementations in Stavanger (and other cases) in depth. This final chapter presents and discusses some main findings from the previous 15 chapters of this book. As such, this chapter draws on the experiences and observations to reflect on Stavanger's path to becoming a smart city.

Within the EU Smart City lighthouse program, Stavanger became a smart city in 2014 (Fernandez et al., 2023) and is currently one of 112 European cities that aim to be climate-neutral by 2030 (Gaglione, 2023). Following Stavanger's history as a resource-dependent city (herring, oil, etc.) and its attempt to become a flagship of green transition and smart city development, like Müller et al. (2023), we now critically reflect on the question, 'What can be learned about the myriad of ways that smart city projects shape and are shaped as they encounter different local, national and Nordic urban development trajectories?'

This chapter provides a synthesis of key findings from the book and reflects on the main lessons learned. We give insights into how the smart sustainable city concept has been implemented and expressed in the city of Stavanger and the everyday life of its inhabitants. The chapter opens with a discussion of Stavanger as a critical case and an 'actual existing smart city' (Shelton et al., 2015). We further reflect critically on opportunities and challenges, in terms of both concrete practical implications and broader theoretical perspectives. The chapter highlights 11 lessons that can be learned from Stavanger, thus summarizing smart city research conducted at the UiS.

DOI: 10.4324/9781003498650-21

Framing

The book includes several case studies, and it is also a case study itself. While case study approaches imply little guidance to the researcher (Thomas, 2011), they are usually united by a commitment to 'studying the complexity that is involved in real situations' (Thomas, 2011, p. 512; cf. Simon, 2009). Researchers may choose a 'local knowledge case' that provides knowledge and ample opportunities for informed, in-depth analysis and discussions, based on the researchers' and practitioners' familiarity with the case. This chapter is based on research conversations and an interpretative process that reveals significant lessons from the book, taking into account complex interrelations, including the authors' perspectives and biases.

(1) Stavanger ... so what?

As a kind of joint 'local knowledge case' (Thomas, 2011, p. 514), many chapters in this book provide insights into the specific urban context of Stavanger and explore smart and sustainable concepts and realities in a medium-sized Nordic smart city. It can therefore be argued that the book does not give a generalizable picture. However, the concrete context of Stavanger and its smart city path has been used as a starting point for exploration, an example of manifestation, or an opportunity for empirical data and research. The single case of Stavanger has provided ample opportunities for informed, in-depth analysis and discussions, based on the researchers' and practitioners' familiarity with the thematic focus and the local context of a lighthouse smart city (cf. Thomas, 2011).

When studying a phenomenon as prolific as the smart city, we believe that experiences from actually existing smart cities like Stavanger, are valuable. As Shelton et al. (2015, p. 14) pointed out, 'rather than the idealised but unrealised vision that often dominates the social imaginary and critique of what a technologically-mediated city might look like', this book allows us to explore the impact of the *smart city* concept through the reality that the researchers and practitioners have observed and experienced in Stavanger or Trondheim, another medium-sized Norwegian city (see Lorentzen and Langhelle, chapter 4). Thus, the book does not assume anticipated outcomes but describes and explores real processes, benefits, and challenges. With the multiple approaches to seeing an existing smart city from various disciplinary points of view (Bibri, 2018), the book contributes to a holistic and in-depth picture of complexities, inspiring further smart city research.

This in-depth study of Stavanger shows different initiatives, methods, and results, ranging from sustainable cycling (Joudavi and Tarigan, chapter 13) and streetwise art experiences (Alberts and Fadnes, chapter 9) and old art recontextualized (Lindland and Matre, chapter 15), to urban planning (Fisker et al., chapter 6, and Müller-Eie, chapter 7) and issues of safe and secure infrastructures and AI-based 5G (Skotnes and Gould, chapter 8, and Nencioni et al., chapter 11). The book elucidates multiple answers to what the smart city concept can mean for a city like Stavanger and gives rich, critical, and possibly contradicting understandings of what smart and sustainable concepts can do to a city and what the city with its inhabitants and initiatives can do to the understanding of the concept. Kunzmann (2010) and Lindtvedt et al. (2021)

underlined the importance of knowledge sharing among international ongoing smart city projects related to medium-sized cities like Stavanger. The case of Stavanger as an early Nordic smart city is thus critical for fostering a more differentiated understanding of smart city reality.

(2) Who says Stavanger is smart and sustainable?

The route to becoming a smart city is often grounded in policy or research projects, as was the case for Stavanger (see Aunemo et al., chapter 1). From there, initiatives grew outwards. This formal introduction of a concept exemplifies how far political will and related funding can go (Stavanger Municipality, 2016). In chapter 10, Undheim and Sageidet elucidate challenges related to the long-term implementation of the smart city concept in Stavanger; inspired by Simon (2009), some may say that Stavanger is about to become smart and sustainable, while others would evaluate differently. For many, Stavanger smart city might just have been a trend in time (see Müller-Eie, chapter 7). Or it may have paved the way for welfare technology and collaboration with the municipal-owned power and telecommunication company (see Syse et al., chapter 12).

The question of what is influenced and achieved through the smart city agenda in Stavanger remains. The physical realities in smart cities are often used as testbeds or living labs for smart and sustainable solutions. These living labs are not confined to physical and technical solutions but provide a means to study how to realize better social conditions for, with, and by the citizens themselves (see Lorentzen and Langhelle, chapter 4). However, it is questionable whether the local population sees Stavanger as a smart city or identifies as living in one (cf. Martin et al., 2019; Øvrebekk, 2019; Paaske, 2017), and it may be challenging to find empirical evidence of what the Triangulum project and Stavanger's smart city office have achieved (Haarstad & Wathne, 2018; Undheim, 2020). However, when viewing the smart city as a political initiative with considerable impact on local roles, structures, collaboration, initiatives, and investment, it can be viewed as a success.

Nevertheless, Stavanger might be an especially challenging city to transform into a smart and sustainable city for two specific reasons. Firstly, Stavanger's welfare and affluence still depend on the oil and gas industry. The smart city strategy therefore needed to create a narrative that re-directs the development into a greener one, while not disregarding the fossil fuel legacy. Secondly, the oil and gas industry background, and especially the municipal-owned power and telecommunication company, have essentially contributed to technological competence in the region (see Aunemo et al., chapter 1), while simultaneously the need for digital solutions to solve public challenges might not be as pressing as in larger cities, because there may be less need to organize and coordinate the public due to fewer social interactions and lower densities in public spaces. It is therefore important to ensure that smart solutions contribute to solving pressing issues, rather than merely testing possible technologies (see Müller-Eie, chapter 7).

While sustainability ambitions have contributed to forming Stavanger as a smart city, the extent of the sustainable aspects and outcomes during the implementation vary. According to Stavanger's environmental plan (Stavanger Municipality, 2018), cycling and the use of public transport should increase, and air quality should improve. However, in chapter 14 Sageidet et al. reveal that there is limited awareness of issues surrounding air pollution, and in chapter 13 Joudavi and Tarigan describe how strategies to promote cycling are somewhat independent of smart initiatives. Furthermore, while last century's artist Kitty Kielland's paintings remind us of the urgency of peatland conversation in the county (see Lindland and Matre, chapter 15), Stavanger's zero CO_2 emissions goal by 2030 is an ambitious orientation towards the Global Sustainable Development Goals (United Nations, 2015) which will require substantial work to succeed. Nevertheless, the ambitious environmental goals of Stavanger and the participation in EU missions may not be in place if it were not for experiences and directions from the smart city initiative.

(3) Out with the old, in with the smart

Another observation is that Stavanger has experienced both continuities and discontinuities (Müller et al., 2023) on its path to a smart city. Initiatives that existed before the smart agenda, particularly within sustainable transport and energy, have continued to exist and have partly been 're-branded' to fall under the umbrella of smart measures. This is a common occurrence in smart city realization (Müller-Eie & Kosmidis, 2023; Parks & Rohracher, 2019), addressed by Müller-Eie in chapter 7 and Joudavi and Tarigan in chapter 13. Given the attention (and funding) that smart measures receive, it is unsurprising that the smart label is used as a crowbar to open certain doors and gain relevance (Haarstad & Wathne, 2018).

This can be viewed similarly to the concept of greenwashing (De Freitas Netto et al., 2020); that is, 'smart-washing' where conventional measures or initiatives are communicated as smart. While one can argue that this contributes to conventional measures reaching their potential under the smart umbrella, it can also mean that smart initiatives receive more attention and thus undermine existing well-functioning and effective measures. On the other hand, as funding and awareness often are related to such buzzwords, it may lead to the re-thinking of previous and existing activities, such as in the example given by Lindland and Matre in chapter 15, where Stavanger Art Museum has used smart art as a way to both create new collaborations but also to explore how old art can help us reconsider the city and its landscape development through environmental lenses.

(4) Pilot the new

Stavanger's pioneer status, especially as one of the first three EU lighthouse cities, has created a ripple effect, making Stavanger a hub for smart city initiatives. Also, UiS has tailored unique study programs and courses to support this focus (see Huang et al., chapter 3). However, when discussing the discontinuities of smart city initiatives, the short lives of pilots and experimental approaches come to mind. As a physical smart city, Stavanger became an arena for urban living labs, understood as pilots for exploring and demonstrating possible solutions in a realistic context (Cardullo et al., 2018), as exemplified by Nencioni et al. in chapter 11. Smart city literature often documents and reports on the testing of smart technology, applications, and processes. This 'demo or die' approach is linked to an understanding that being able to pilot or demonstrate an intervention under certain conditions makes it more realistic and reliable in the future (Halpern & Günel, 2017), thus assuming generalizability and scalability. Such experimentation can develop positive urban development as in the case of Svartlamoen in Trondheim described by Lorentzen and Langhelle in chapter 4 or reveal challenging starting points for digital development as described by Nencioni et al. in chapter 11.

However, there are several aspects to consider about pilots and demonstrations. On one hand, the small urban testbeds or pilot projects often cannot readily scope all environmental, social, and economic uncertainties that can occur in the future. This weakens the validity of such approaches and may contribute to overestimating the effect certain initiatives can have. On the other hand, after a pilot is terminated, its continuation is often not part of the existing strategy or budget and thus is not accounted for (see Undheim and Sageidet, chapter 10). In the case of successful pilots, good initiatives may be discontinued despite their potential positive impact. It seems that an agenda to continue successful interventions after a trial period is lacking. In the case of failure, pilots should be sufficiently reported to increase the learning potential. Failures often have far more learning points than successes because they clarify whether pre-existing assumptions do not align with the experienced reality (see Nencioni et al., chapter 11). Further, the lack of negative results can lead to possible re-investment in similar initiatives elsewhere or later.

(5) Co-creating smartness

Methods for how to involve citizens in a democratic and just way and how to do this effectively are emphasized in smart city approaches (Paskaleva et al., 2015) and have opened local authorities as well as entrepreneurs to new ways of thinking about participatory design and co-creation as shown by Undheim and Sageidet in chapter 10. In Stavanger, this has been particularly prolific and led to new university education in co-creation and design thinking. However, it is often middle-class citizens of the ethnic and cultural majority who are engaged in co-creation community activities. This may lead to an over-emphasis on the 'imagined average' and exclusion of marginalized and vulnerable groups, as depicted by Fisker et al. in chapter 6. As citizens and stakeholders become increasingly familiar with co-creation activities, a greater awareness and acknowledgement of the importance of participatory processes have emerged and been widened towards global and environmental citizenship, including children's agency (see Sageidet et al., chapter 14). However, the activities are in danger of becoming scripts for legitimizing urban development without necessarily leading to more democratic involvement (Montalvan Castilla & Riel Müller, 2023). The question is whether the co-creation activities can survive long-term unless the citizens themselves raise the agenda and take ownership, as was the case in Svartlamoen (see Lorentzen and Langhelle, chapter 4). Succeeding in developing better urban solutions requires the active involvement of those affected. However, figuring out how to engage in a participative and just way is still challenging. As we depend on the insight and contributions of those affected by the developments, socially just co-creation and co-production and related methods and processes should not be given up on.

(6) Smart and sustainable? Prove it!

As described above, smart initiatives are often driven by whether they work or are cost-effective. Here, a demonstration project or a pilot can give valuable insight. Conducting certain projects as pilots also releases stakeholders from long bureaucratic processes and thorough impact analyses. Furthermore, such initiatives can sometimes be implemented without explicitly stating goals or measurable effects (Müller-Eie & Kosmidis, 2023). While Lindland suggests in chapter 5 that impact is a continuous process, and thus not a final and static result, Müller-Eie explains in chapter 7 how the uncertainty of effects undermines urban development approaches and investment by removing accountability from projects. However, only when initiatives are rolled out on a larger scale or over time do we start to ask what the societal impacts are. Theoretically, the smart city is a city of open big data sourcing and use (Bibri, 2018) and thus democratic in sharing its information and data. Hypothetically, this enables independent analysis and provides measurability and accountability. However, research efforts into energy consumption such as those by Syse et al. in chapter 12 have shown that data access is challenging due to gatekeeping and privacy regulations. Another challenge is that societal impact, such as changes in attitudes and patterns of behaviour, only become visible over time and are not identifiable here and now. Then the ontological question arises of whether something that is not measurable can exist.

(7) I say smart, you say smart ...

The smart city of Stavanger has been studied from different perspectives. Researchers have looked at how different elements can contribute to achieving various goals; for example, how certain technologies can contribute to improving the organizing of sustainable mobility and motivating people to cycle (see Joudavi and Tarigan, chapter 13) as theorized in Winkowska et al. (2019). Other times the aim has been to involve citizens in the development of a new city area to cater towards social inclusion and sustainability (Fisker et al., chapter 6, and Sageidet et al., chapter 14). Other projects aim at developing or identifying the right process for a certain challenge, implying a more substantive understanding of the topic (see Nencioni et al., chapter 11). How different disciplines define and conceptualize smart city concepts is interesting as it strongly influences how we discuss and approach the concept and its relevance. Maybe we need a shared smart city definition, or we need a better understanding of what a smart city can mean in different contexts (Albino et al., 2015; Keshavarzi et al., 2021).

This becomes apparent when data science treats the smart city as a goal (see Nencioni et al., chapter 11), whereas urban planning views the smart city as a means to achieve sustainable urban development and mobility (see Müller-Eie, chapter 7). Yet another understanding is found in chapter 5, where Lindland, representing the social sciences, views the smart city concept as a social and dynamic meaning-bearing term under continuous development. At best, this can cater towards a widening and deepening, and thus enhancing, of the smart city concept, what it is, what it can be, and how to create synergetic effects when implementing it. However, it may also lead to misunderstanding, false consensus, or disagreement among collaborating partners (Gundersen & Jacobsen, 2023). Viewing the smart city concept and related discourses through different disciplinary lenses can cater to a wider and more appropriate understanding of smart cities. It is also important to cater towards more intellectual exchange between different fields and disciplines as demonstrated by Huang et al. in chapter 3. Because the cross-disciplinary variations contribute to different interpretations of the same situations, they open the way for novel understandings that are not possible to reach within a single perspective.

(8) What about us?

Building on this, there seem to be distinct strands of smart city research. On the one hand, the origin of the smart city concept is grounded in information and communications technology as a driver for urban efficiency and sustainability. This strand includes solutions with a techno-centric corporate smart city model, mainly applying to 'hard' domains such as buildings, energy grids, water, waste, and resource management, mobility, and logistics. However, there is a general lack of critical reflection on how urban development rooted in technological advancement and entrepreneurial thinking can impact the very nature of cities and urban life (Kitchin, 2015; Komninos et al., 2019). On the other hand, there is a call for a more holistic smart city view including 'soft' domains such as policy, governance, culture, and education (Albino et al., 2015; Mora et al., 2017). In Stavanger, early collaborations with UiS have been predominantly technological, while the more holistic view has unfolded in conjunction with the development of the Nordic smart city model based on Nordic social values (DOGA et al., 2019; Kangas & Kvist, 2019), as laid out by Undheim and Sageidet in chapter 10.

Since the inception of the smart city, Stavanger has included less conventional subjects such as smart art, evidenced by Alberts and Fadnes in chapter 9 and Lindland and Matre in chapter 15. As elaborated by Müller in chapter 2, the Research Network for Smart Sustainable Cities at the UiS has also chosen an approach to smart cities that is rather broad and thus reached unconventional disciplines working with smart and sustainable development, such as early childhood education (see Sageidet et al., chapter 14). This inclusive approach to the smart city can lead to new insights and answer the call for different disciplines, fields, and research communities to integrate efforts in smart city research to effectively counteract a growing divide (Mora et al., 2017). Including 'hard' and 'soft' aspects of smart cities caters towards a stronger integration of these different spheres.

(9) Collaboration. Collaboration. Collaboration.

Along similar lines, the smart city must be understood as an inherently inter-, multi-, and cross-disciplinary endeavour, and possibly as a transdisciplinary concept according to Huang et al. (chapter 3). The only way to address this is by keeping something as oxymoronic as a holistic focus. Internationally, there has been an evolution from merely technology-focused approaches to the smart city towards a more holistic perspective, integrating governance, social capital, urban development, people, and community needs (Albino et al., 2015). At UiS, the importance of interdisciplinary work within energy, sustainability, health, and technology is especially emphasized.

In Stavanger the triple helix and quadruple helix (Cai & Lattu, 2021) have become common collaborative concepts when working with the smart city (see Aunemo et al., chapter 1). This has spread to other urban development approaches, where local authorities now have a greater understanding of the value of research-based innovation and entrepreneurial spirit in moving the city forward. However, true multi-, cross-, trans-, and interdisciplinary collaboration requires not only institutions to collaborate but also single organizations to offer institutional incentives catering to interdepartmental research and discourse, as pointed out by Huang et al. in chapter 3. They also call for facilitation by raising awareness of the importance of collaborating, as well as having good structures and incentives in place to do so and, not least, time. The establishment of well-founded collaborations requires trust and engagement, which take time to establish. Interdisciplinary cooperation is pivotal, with knowledge sharing as a cornerstone.

(10) Everybody's an entrepreneur all of a sudden

The smart city vision for Stavanger has been driven by key persons from academia and practice. The smart city vision also required public authorities to adjust the way they understand their own roles. This is evident in the staffing of strategic positions in the municipality and county, where now more people with a background in technology and business are hired instead of technocrats or bureaucrats. In chapter 2 Müller highlights how persons with the right drive, network, and collaboration skills have had key roles in Stavanger, both during the process of getting funding from the EU (see Aunemo et al., chapter 1) and during the implementation and maintenance of smart urban concepts and practices. In chapter 4, Lorentzen and Langhelle point out the key roles that certain organizations and people hold in building bridges and negotiating between different stakeholders. Such enthusiasts are often important resources in these innovative processes.

This also implies that the smart city concept comes with a philosophy and ontology embedded in contractual thinking, rather than a local community feeling – or as rooted in *Gesellschaft* rather than *Gemeinschaft* (Tönnies & Loomis, 1999). Economic growth and private market initiatives are seen as imperative for the position of Europe in the world, and the marked-based entrepreneurship inherent to the smart city concept aligns well with this need.

(11) Smart to be sustainable? Sustainable to be smart?

There are essential interrelations between smart cities and sustainable development, defined by the Brundtland report (World Commission on Environment and Development, 1987) and the Sustainable Development Goals (United Nations, 2015), acknowledged by the Roadmap for Norwegian Smart and Sustainable Cities (DOGA et al., 2019), and currently promoted through the EU's climate-neutral and smart cities (Gaglione, 2023). Ahvenniemi et al. (2017) noted that there has been a shift from assessing urban performance according to sustainability indicators towards smart city goals, and Karvonen et al. (2020) stated that the smart city competes with and sometimes overshadows sustainable urban development agendas. Roggema (2020) revealed that 'smart' and 'sustainable' are used dichotomously and that research into these concepts seems divided. Therefore, multiple authors have called for a clearer integration or delineation between the concepts (Cugurullo, 2018; Karvonen et al., 2020; Roggema, 2020). Since it is still unclear how the smart city concept relates to urban sustainable development (Haarstad, 2017), it is critical to understand in which ways smart projects or strategies can impact urban life, urban behaviours, and urban societies at large (Khansari et al., 2014).

While Stavanger's strategies of smart and sustainable development have occurred hand in hand, it is unclear how they relate to each other (Haarstad, 2017). We have also tried to show how they can be aligned in some situations and contrast one another in other situations (see Müller-Eie, chapter 7, and Nencioni et al., chapter 11). Smartness and sustainability, as well as their relationship, can also change over time, and several views on their interrelation may exist simultaneously. It will be interesting to observe how the strategic measures to become climate-neutral overlap with existing or previous smart measures in Stavanger. It is possible that Stavanger has the potential to become a frontrunner in the green transition and can serve as a powerful example for other cities and municipalities worldwide. However, that remains to be seen.

What we learned from Stavanger

In summary, we can conclude that the Research Network for Smart Sustainable Cities at the UiS has generated research that gives rise to several lessons to be learned from the case of Stavanger as a critical example of a Nordic smart city.

Lesson 1: In-depth context-dependent knowledge from studying the smart city case of Stavanger reveals complexities and nuances and provides valuable and transferable knowledge.
Lesson 2: Translating policy into reality is a powerful way of showing political will, but it takes a long time to fundamentally embed new concepts into a city and its identity.
Lesson 3: The smart label can sometimes be used to promote conventional approaches, leading to 'smart' becoming an all-encompassing term void of

meaning or contributing to reconsidering current ideas and new interpretations and collaborations.

Lesson 4: Experimental pilots and demonstrations can be short-lived unless their continuation is accounted for in case of success or their discontinuation is well documented and used for exploration of alternative understandings in case of 'failure'.

Lesson 5: Awareness of participatory design and co-creation has increased and infiltrated other processes.

Lesson 6: Smart and sustainable results can be difficult to measure or test for cost-effectiveness due to a lack of data access or integrated plans for impact analysis.

Lesson 7: Creating a multidisciplinary understanding of the smart city requires good communication, understanding of varied goals and measures, and mutual respect and humility.

Lesson 8: An inclusive interpretation of a smart sustainable city caters to creating new, more exploratory, and alternative perspectives and insights into what smart urban development can be.

Lesson 9: Collaboration between different stakeholders, sectors, and disciplines is essential for innovative approaches to smart sustainable development. It takes knowledge, structures, resources, and time to develop and maintain.

Lesson 10: Smart and sustainable development requires innovative thinking, and the public sector assimilates entrepreneurial strategies to realize its vision. This often includes the engagement of enthusiasts in key positions.

Lesson 11: Being explicit about the differences and similarities between smart and sustainable is important and helps to address the achievement of both.

The above lessons, derived from Stavanger as an existing city, hold relevant insights into the reality of smart cities as studied by researchers and experienced by the local population. As such, the book contributes to a multi-faceted development of the smart city concept. The lessons are generalizable, transferable, and applicable in other contexts. As a contribution to a growing body of empirical insight into 'actually existing smart cities', and with critical reflections on the societal impact of smart initiatives, the book caters to a more nuanced understanding of barriers to fully integrating smart cities and sustainable urban development. Further, this constitutes an attempt to view and explore one specific area from a multitude of different disciplines, to expand the smart city concepts to include less conventional disciplines, such as art/music and early childhood education. These contributions benefit the research community and practitioners concerned with the smart city by offering insights and opening avenues to yet-unexplored facets of the *smart city* as a concept and a lived reality.

References

Ahvenniemi, H., Huovila, A., Pinto-Seppä, I., & Airaksinen, M. (2017). What are the differences between sustainable and smart cities? *Cities*, 60, 234–245. doi:10.1016/j.cities.2016.09.009

Albino, V., Berardi, U., & Dangelico, R.M. (2015). Smart cities: Definitions, dimensions, performance, and initiatives. *Journal of Urban Technology*, 22(1), 3–21. doi:10.1080/10630732.2014.942092

Bibri, S.E. (2018). A foundational framework for smart sustainable city development: Theoretical, disciplinary, and discursive dimensions and their synergies. *Sustainable Cities and Society*, 38, 758–794. doi:10.1016/j.scs.2017.12.032

Cai, Y., & Lattu, A. (2021). Triple helix or quadruple helix: Which model of innovation to choose for empirical studies? *Minerva*, 60, 257–280. https://link.springer.com/article/10.1007/s11024-021-09453-6

Cardullo, P., De Feliciantonio, C., & Kitchin, R. (2018). *The right to the smart city*. Emerald. doi:10.1108/9781787691391

Cugurullo, F. (2018). Exposing smart cities and eco-cities: Frankenstein urbanism and the sustainability challenges of the experimental city. *Environment and Planning A: Economy and Space*, 50(1), 73–92. doi:10.1177/0308518X17738535

De Freitas Netto, S.V., Sobral, M.F.F., Ribeiro, A.R.B., & Soares, G.R.D.L. (2020). Concepts and forms of greenwashing: A systematic review. *Environmental Sciences Europe*, 32(1). doi:10.1186/s12302-020-0300-3

DOGA, Norwegian Smart City Network, & Nordic Edge. (2019). *Roadmap for smart and sustainable cities and communities in Norway. A guide for local and regional authorities developed by Design and Architecture Norway (DOGA), the Norwegian Smart City Network and Nordic Edge.* https://doga.no/globalassets/pdf/smartby-veikart-19x23cm-eng-v1_delt.pdf

Fernandez, T., Stöffler, S., & Diaz, C. (2023). 3-Triangulum: The three point project – Findings from one of the first EU smart city projects. In P. Droege (Ed.), *Intelligent environments* (2nd ed., pp. 87–107). Elsevier. doi:10.1016/B978-0-12-820247-0.00010-2

Gaglione, F. (2023). Policies and practices of transition towards climate-neutral and smart cities. *TeMA - Journal of Land Use, Mobility and Environment*, 16(1), 227–231. doi:10.6093/1970-9870/9822

Gundersen, S.R., & Jacobsen, R.H. (2023). *Samskapingsprosessen i pilotprosjektet NEB-STAR* [Master's thesis, University of Stavanger]. https://uis.brage.unit.no/uis-xmlui/handle/11250/3086065

Haarstad, H. (2017). Constructing the sustainable city: Examining the role of sustainability in the 'smart city' discourse. *Journal of Environmental Policy & Planning*, 19(4), 423–437. doi:10.1080/1523908X.2016.1245610

Haarstad, H., & Wathne, M. W. (2018). Smart cities as strategic actors: Insights from EU Lighthouse projects in Stavanger, Stockholm, and Nottingham. In I.A. Karvonen, F. Cugurullo, & F. Caprotti (Eds.), *Inside smart cities: Place, politics and urban innovation* (p. 24). Routledge.

Halpern, O., & Günel, G. (2017). *Demoing unto death: Smart cities, environment, and preemptive hope*. Twentynine.

Kangas, O., & Kvist, J. (2019). Nordic welfare states. In B. Greve (Ed.), *Routledge handbook of the welfare state* (2nd ed.). Routledge.

Karvonen, A., Cook, M., & Haarstad, H. (2020). Urban planning and the smart city: Projects, practices and politics. *Urban Planning*, 5(1), 65–68. doi:10.17645/up.v5i1.2936

Keshavarzi, G., Yildirim, Y., & Arefi, M. (2021). Does scale matter? An overview over the 'smart city' literature. *Sustainable Cities and Societies*, 74, 103151. doi:10.1016/j.scs.2021.103151

Khansari, N., Mostashari, A., & Mansouri, M. (2014). Impacting sustainable behavior and planning in smart city. *International Journal of Sustainable Land Use and Urban Planning*, 1(2), 46–61. doi:10.24102/IJSLUP.v1I2.365

Kitchin, R. (2015). Making sense of smart cities: Addressing present shortcomings. *Cambridge Journal of Regions, Economy and Society*, 8(1), 131–136. doi:10.1093/cjres/rsu027

Komninos, N., Kakderi, C., Panori, A., & Tsarchopoulos, P. (2019). Smart city planning from an evolutionary perspective. *Journal of Urban Technology*, 26(2), 3–20. doi:10.1080/10630732.2018.1485368

Kunzmann, K.R. (2010). Medium-sized towns, strategic planning, and creative governance. In M. Cerrata, G. Concilio, & V. Monno (Eds.), *Urban and landscape perspectives: Making strategies in spatial planning* (Vol. 9, pp. 27–45). Springer. doi:10.1007/978-90-481-3106-8_2

Lindtvedt, I.C.R., Frøhaug, R.S., & Nesse, P.J. (2021). Smart city development in Nordic medium-sized municipalities. *Nordic and Baltic Journal of Information and Communications Technologies*, 2021, 1–48. https://doi.org.10.13052/nbjict1902-09-097Xy.2021.001

Martin, C., Evans, J., Karvonen, A., Paskaleva, K., Yang, D., & Linjordet, T. (2019). Smart-sustainability: A new urban fix? *Sustainable Cities and Society*, 45, 640–648. doi:10.1016/j.scs.2018.11.028

Montalvan Castilla, J.E., & Riel Müller, A. (2023). A smart city for all citizens: An exploration of children's participation in Norway's smartest city. *International Planning Studies*, 29(1), 19–33. doi:10.1080/13563475.2023.2259110

Mora, L., Bolici, R., & Deakin, M. (2017). The first two decades of smart-city research: A bibliometric analysis. *Journal of Urban Technology*, 24(1), 3–27. doi:10.1080/10630732.2017.1285123

Müller, A.R., Park, J., & Sonn, J.W. (2023). Finding the old in the new: Smart cities in the national and local trajectories of urban development. *International Journal of Urban Sciences*, 27, 1–9.

Müller-Eie, D., & Kosmidis, I. (2023). Sustainable mobility in smart cities: A document study of mobility initiatives of mid-sized Nordic smart cities. *European Transport Research Review*, 15(1), 36. doi:10.1186/s12544-023-00610-4

Øvrebekk, H. (2019, September 24). *Ut med teknologi, inn med mennesker.* aftenbladet. no. https://www.aftenbladet.no/meninger/kommentar/i/QoOXLR/ut-med-teknologi-inn-med-mennesker

Paaske, C. (2017, October 24). *Er smartby orwellsk nytale for at vi egentlig kommer til å bli dummere?*aftenbladet.no. https://www.aftenbladet.no/meninger/debatt/i/6r4z3/smartby-en-by-av-roboter

Parks, D., & Rohracher, H. (2019). From sustainable to smart: Re-branding or re-assembling urban energy infrastructure? *Geoforum*, 100, 51–59.

Paskaleva, K., Cooper, I., Linde, P., Peterson, B., & Götz, C. (2015). Stakeholder engagement in the smart city: Making living labs work. In M. Rodríguez-Bolívar (Ed.), *Transforming city governments for successful smart cities* (pp. 115–145). Springer. doi:10.1007/978-3-319-03167-5_7

Roggema, R. (2020). Towards integration of smart and sustainable cities. In R. Roggema & A. Roggema (Eds.), *Smart and sustainable cities and buildings* (pp. 5–23). Springer. doi:10.1007/978-3-030-37635-2

Shelton, T., Zook, M., & Wiig, A. (2015). The 'actually existing smart city'. *Cambridge Journal of Regions, Economy and Society*, 8(1), 13–25, doi:10.1093/cjres/rsu026

Simon, H. (2009). *Case study research in practice.* Sage.

Stavanger Municipality. (2016). *Veikart for smartbyen Stavanger – Visjon, mål og satsingsområder.* https://issuu.com/stavanger.kommune/docs/veikart_for_smartbyen_stavanger_121/1?ff&e=4979314/53937755

Stavanger Municipality. (2018). *Klima- og milljøplan.* https://www.stavanger.kommune.no/siteassets/renovasjon-klima-og-miljo/miljo-og-klima/klima–og-miljoplan-2018-2030.pdf

Thomas, G. (2011). A typology for the case study in social science following a review of definition, discourse, and structure. *Qualitative Inquiry*, 17(6), 511–521. doi:10.1177/1077800411409884

Tönnies, F., & Loomis, C.P. (1999). *Community and society* (1st ed.). Routledge. (Original work published 1887)

Undheim, S. (2020). Implementation of the smart city concept in Stavanger Municipality: A case study of how Smart City Stavanger and selected measures meet the field of practice [Master's thesis, University of Stavanger]. https://uis.brage.unit.no/uis-xmlui/handle/11250/2682071

United Nations. (2015). *Transforming our world: The 2030 agenda for sustainable development*. https://sdgs.un.org/2030agenda

Winkowska, J., Szpilko, D., & Pejić, S. (2019). Smart city concept in the light of the literature review. *Engineering Management in Production and Services*, 11(2), 70–86. doi:10.2478/emj-2019-0012

World Commission on Environment and Development. (1987). *Our common future. A report from the United Nations World Commission on Environment and Development.* Oxford University Press.

Index